The Future of Genetically Modified Crops

Lessons from the Green Revolution

FELICIA WU

WILLIAM P. BUTZ

RAND SCIENCE AND TECHNOLOGY

This research in the public interest was supported by RAND, using discretionary funds made possible by the generosity of RAND's donors and the fees earned on client-funded research.

Library of Congress Cataloging-in-Publication Data

Wu, Felicia.
  The future of genetically modified crops : lessons from the Green Revolution / Felicia Wu and William Butz.
    p. cm.
    "MG-161."
    Includes bibliographical references.
    ISBN 0-8330-3646-7 (pbk.)
    1. Transgenic plants. 2. Crops—Genetic engineering. 3. Green revolution.
  I. Butz, William P. II. Title.

SB123.57.W8 2004
631.5'233—dc22

                                                                    2004014614

The RAND Corporation is a nonprofit research organization providing objective analysis and effective solutions that address the challenges facing the public and private sectors around the world. RAND's publications do not necessarily reflect the opinions of its research clients and sponsors.

RAND® is a registered trademark.

Cover design by Peter Soriano

Published 2004 by the RAND Corporation
1700 Main Street, P.O. Box 2138, Santa Monica, CA 90407-2138
1200 South Hayes Street, Arlington, VA 22202-5050
201 North Craig Street, Suite 202, Pittsburgh, PA 15213-1516
RAND URL: http://www.rand.org/
To order RAND documents or to obtain additional information, contact
Distribution Services: Telephone: (310) 451-7002;
Fax: (310) 451-6915; Email: order@rand.org

Preface

The number of people in danger of malnutrition worldwide has decreased significantly in the past 30 years, thanks in part to the Green Revolution of the 20th century. However, an estimated 800 million people still lack adequate access to food. The world now sits at the cusp of a second potential agricultural revolution, the "Gene Revolution" in which modern biotechnology enables the production of genetically modified (GM) crops that may be tailored to address ongoing agricultural problems in specific regions of the world. The GM crop movement has the potential to do enormous good, but also presents novel risks and has significant challenges to overcome before it can truly be considered revolutionary. This monograph seeks to answer these questions: Can the Gene Revolution become in fact a global revolution, and, if so, how should it best proceed?

This report draws on lessons from the Green Revolution to inform stakeholders who are concerned with the current GM crop movement. We hope that this analysis can illuminate opportunities for GM crops to increase farm production, rural income, and food security in developing countries, while controlling potential risks to health and the environment. The analysis and findings in this report are intended for all individuals and institutions interested in improving agricultural production and food quality in the developing world, and particularly those who have a stake in the worldwide debate over genetically modified crops.

This report results from the RAND Corporation's continuing program of self-sponsored independent research. Support for such research is provided, in part, by donors and by the independent research and development provisions of RAND's contracts for the operation of its U.S. Department of Defense federally funded research and development centers.

Questions about this report should be directed to Felicia Wu at the University of Pittsburgh, Graduate School of Public Health, A718 Crabtree Hall, 130 DeSoto St., Pittsburgh, PA 15261 (fwu@ eoh.pitt.edu).

The RAND Corporation Quality Assurance Process

Peer review is an integral part of all RAND research projects. Prior to publication, this document, as with all documents in the RAND monograph series, was subject to a quality assurance process to ensure that the research meets several standards, including the following: The problem is well formulated; the research approach is well designed and well executed; the data and assumptions are sound; the findings are useful and advance knowledge; the implications and recommendations follow logically from the findings and are explained thoroughly; the documentation is accurate, understandable, cogent, and temperate in tone; the research demonstrates understanding of related previous studies; and the research is relevant, objective, independent, and balanced. Peer review is conducted by research professionals who were not members of the project team.

RAND routinely reviews and refines its quality assurance process and also conducts periodic external and internal reviews of the quality of its body of work. For additional details regarding the RAND quality assurance process, visit http://www.rand.org/standards/.

Contents

Figures

Tables

Summary

The world now sits at the cusp of a new agricultural revolution—the "Gene Revolution" in which modern biotechnology enables the production of genetically modified (GM) crops that may be tailored to address agricultural problems worldwide. This report investigates the circumstances and processes that can induce and sustain such an agricultural revolution. It does so by comparing the current GM crop movement with the Green Revolution of the latter half of the 20th century. We assess not only the scientific and technological differences in crops and in agricultural methods between these two movements, but more generally the economic, cultural, and political factors that influence whether a new agricultural technology is adopted and accepted by farmers, consumers, and governments. Our historical analysis of the earlier Green Revolution provides lessons about whether and how genetically modified crops might spread around the world. Whether the latter movement will develop into a global Gene Revolution remains to be seen.

Genetically modified crops created by modern agricultural biotechnology have attracted worldwide attention in the past decade. Cautious voices warn that the health and environmental effects of GM crops are uncertain and that their cultivation could have unintended adverse consequences. Alternatively, supporters of the technology assert that GM crops could revolutionize world agriculture, particularly in developing countries, in ways that would substantially reduce malnutrition, improve food security, and increase rural income, and in some cases even reduce environmental pollutants.

Can the GM crop movement develop into an agricultural revolution on the scale of the Green Revolution? To answer this question, first, it is important to consider what an agricultural revolution entails. Viewed historically, movements that come to be considered agricultural revolutions share the following features:

1. The movements gave farmers incentives to produce—i.e., the technologies provided a net benefit to farmers.
2. The movements substantially improved agricultural production, food nutrition, or both; or they substantially decreased necessary inputs such as fertilizer or water.
3. People were generally willing to adapt culturally and economically to the new technologies, and consumers accepted the products of the agricultural movement.
4. There was cooperation among those that provided the technologies, regulated the technologies, and used the technologies.
5. The movements were sustainable, eventually without public subsidization.

On a regional scale, GM crops might indeed be considered revolutionary—that is, they could meet all five criteria for an agricultural revolution. In the United States, Canada, China, and Argentina, for example, genetically modified varieties of soybeans, corn, and cotton now make up from about a third to 80 percent of total plantings of those crops, and provide benefits for growers such that these GM varieties will likely continue to make up a substantial portion of total plantings in the foreseeable future. Likewise, policymakers and the general public in these nations are accepting of this new technology. Adoption of these GM crops has led to improved yield, decreased use of pesticides or particularly harmful herbicides, and, in some cases, improved food quality.

While farmers in other nations, such as India and South Africa, have more recently begun to plant GM crops and experience the beginnings of a potential Gene Revolution, the revolution has yet to occur on a global scale. It has stalled because consumer and environmental concerns, along with precautionary regulations, have limited its spreading to the countries that could benefit from it most, notably much of sub-Saharan Africa where famine continually threatens the population.

As stated above, the purpose of this report is to better understand whether and how this GM movement might become an authentic agricultural revolution by comparing it with an earlier agricultural movement that did reach nearly the entire world. The Green Revolution that had its origins in the 1940s, and reached its peak in the 1970s, continues to affect agricultural practices today. By analyzing the Green Revolution's objectives, science and technology, sources of financing, regulatory environment, and ultimate successes and failures, we offer an assessment of the ongoing GM crop movement—whether and how it might make a revolutionary impact on world agriculture.

The stated objective of the Green Revolution was to increase food production in regions of the world facing impending massive malnutrition. In the post-World War II era, scientists and policymakers considered those regions to be Latin America and Asia. Some argue, in retrospect, that this geographic choice was also motivated by Cold War politics: a largely U.S.-supported effort to prevent the spread of communism by ensuring adequate food supplies in at-risk countries.

Regardless of its motivation, the introduction of high-yield varieties (HYVs) of crop seed, along with pesticides, fertilizers, and irrigation systems, transformed agriculture on those two continents. With initial funding from the Rockefeller Foundation, individuals including U.S. plant breeders, agronomists, entomologists, soil scientists, and engineers worked in developing nations while training local agricultural scientists to extend the work in their own locales. The World Bank, Food and Agricultural Organization of the United Nations (FAO), United States Agency for International Development (USAID), and other national and international organizations later joined the Rockefeller Foundation to make this effort succeed. And succeed it did, in terms of increasing food production in Asia, Latin America, and even parts of the industrialized world such as Great

Britain. In Africa, however, where the movement came later, the Green Revolution has yet to improve food production in a sustainable way. As such, this movement provides several important lessons for understanding the possible course of the Gene Revolution.

We compare the Green Revolution and the current GM crop movement in four basic areas: science and technology, funding sources, where the movement occurred or is occurring, and the policies and political motivations surrounding each movement.

Science and Technology

The Green Revolution presented a considerable advance in agricultural technologies for farmers in the developing world, and, to a limited extent, in industrialized countries as well. For the first time, scientists and plant breeders integrated their research with farming practices in traditional agriculture to tackle problems that were constraining crop yield. High-yield seeds for rice, wheat, and corn were introduced in parts of the world where these crops made up a significant portion of the daily diet, and subsequently of food exports. Pesticides, chemical fertilizers, and irrigation systems were also introduced to aid farmers in controlling previously unmanageable pests, dealing with low-quality soil, and delivering water to crops according to their requirements.

The Gene Revolution, propelled by genetic engineering, allows previously unheard-of combinations of traits across species to achieve pre-specified objectives. For example, daffodil and bacterial genes can be introduced into the rice genome so that the rice produces beta-carotene, the precursor of vitamin A. The benefits of the current varieties of GM crops include yield increase, reduced agricultural inputs such as pesticides and fertilizers, reduced vulnerability to the whims of nature, and improved nutritional content. For the most part, these benefits have been limited to parts of the industrialized world to which current GM crop development and marketing have been targeted and, among those, to countries that have allowed their cultivation. Other GM crops are now being developed that survive on less water, that survive in soil heavy in salt or metals such as aluminum, that convert or "fix" nitrogen from the air, and that produce vaccines against common diseases such as cholera and hepatitis B (Byrne et al., 2004).

A fundamental challenge in this newest agricultural movement that did not arise during the Green Revolution is the definition and treatment of intellectual property (IP). IP issues are central to the Gene Revolution because whereas science and technology move forward through the sharing of ideas and resources, IP ambiguities and restrictions can often limit the valuable diffusion of science and technology. Commercial application of biotechnology has taken place primarily in the United States and primarily through the private sector. The issue of who "owns" a particular *event* (the successful transformation) of a genetically modified crop and who can develop it further has become so economically important and contentious that numerous cases involving this issue are being litigated (Woodward, 2003). Some observers consider IP issues to be among the most important impediments to the development and adoption of GM crops in the developing world (Shoemaker et al., 2001; Cayford, 2004). Patent rights that universities may have on their sponsored research, corporate profit interests, and the ability of farmers to buy IP-protected seed are salient IP issues.

Funding

Philanthropic organizations, i.e., the Rockefeller and Ford Foundations, provided the backbone of early funding for the Green Revolution (Perkins, 1997; Pinstrup-Andersen and Schioler, 2001). The scientists who created high-yield seeds and their associated pesticides and fertilizers worked in conjunction with, and were funded by, these foundations along with the governments of Mexico, India, and several other countries. In 1971, while the Green Revolution was bearing its first fruits in many parts of the world, the Consultative Group on International Agricultural Research (CGIAR)—a system of 16 Future Harvest Centers working in more than 100 countries—was created. With the creation of CGIAR, support for developing world agriculture became more broad-based and included European nations, Canada, and Japan.

Genetically modified crops are largely the product of private industry. This is partly because new technologies are far more costly than existing ones, and the biotechnology industry was able to gather the necessary funds to develop these technologies long before public awareness of GM crops could lead to publicly generated funding for GM crop development (Pinstrup-Andersen and Schioler, 2001). Successful companies typically focus on their markets with the intent of generating profit. With regard to agricultural biotechnology, companies in the United States and elsewhere have thus far created primarily seeds that farmers in industrialized countries can and will purchase: corn and soybeans that can tolerate a particular herbicide, corn and cotton that are resistant to particular pests, and food crops that last longer on the supermarket shelf. Because of the "technology fee" that growers pay to use these crop seeds (including recoupment of industry's research and development costs as well as profit), and because the seeds are designed particularly for their planting situations, the targeted farmers in industrial countries have generally found it worthwhile to buy these seeds and have been willing to pay the technology fee (Wu, 2004). Thus, in industrialized nations, GM crop technology has had the potential to revolutionize farming. However, the current GM crop seed varieties are neither affordable nor useful to most of the poorer farmers in the world; hence, their revolutionary impact in the developing world has been limited thus far. Indeed, there seems to be a mismatch of setting and technology, due to the funding sources of basic research.

Some agricultural biotechnology companies have recently expressed interest in working with regional research institutions to develop crops that would be profitable and affordable for farmers in developing countries. In addition, they are willing to donate a substantial portion of their scientific knowledge, such as genomes of key food crops, to increase agricultural knowledge in the developing world. In this way, the challenges related to IP may be lessened.

Where the Revolution Was and Is Taking Place

The Green Revolution was a success, in terms of its stated objectives, in Mexico and the rest of Latin America, India, and much of Southeast Asia. On the other hand, the Green Revolution has had little significant impact in most areas of Africa. Two prominent hypotheses for this outcome are that the technology package that was so useful in some parts of the world was not applicable to African farms, and that rural transportation systems are ill-designed to deliver either the technologies or their resulting products.

The technologies introduced in Asia and Latin America in the Green Revolution generally required not more land, but chemical fertilizer and well-timed water. Farmers who could access these inputs did well while others did not. To the extent that large landholders also had access to fertilizer and irrigation, they tended to adopt the new technologies early and successfully.

It may be too early to predict the varying adoption rates and benefits of yet undeveloped Gene Revolution technologies given the differing characteristics of farmers and regions. What can be said from the Green Revolution experience is that farmers will not adopt and utilize technologies over the long term that do not cost considerably less than current technologies, produce considerably more than current technologies, or substantially reduce the variability of cost or production in their own locales. As opposed to the Green Revolution, the key component of the Gene Revolution technology is improved seed. This being the case, all farmers, small or large, should be able to take advantage of the Gene Revolution; theoretically, the Gene Revolution is scale-neutral, providing that one can pay for the seed. However, cultural factors may deter farmers from embracing the new science; genetically modified crops have already become a stigmatized technology in some parts of the world because of concerns about manipulating organisms in seemingly "unnatural" ways and fears of unintended adverse impacts on the environment or human health.

Policies and Politics

At the time the Green Revolution was first seriously considered, the United States and the rest of the developed world feared that food crises in developing countries would cause political instability that could push those countries over to the Communist side (Perkins, 1997). Partly as a result of this issue, the U.S. government was highly concerned about agricultural science in the developing world and worked with foundations and scientists in the post-World War II decades to bring about the Green Revolution in regions subject to famine.

As of yet, there does not appear to be a strong political motivation for genetically modified crops to succeed in the developing world. Communism is no longer a threat, and famines, while still a problem in parts of the world, appear to be more the result of localized weather, politics, and war conditions than a sweeping threat that commands sustained government and public attention in industrial countries. Instead, public concerns and national and international regulations are now the driving force behind whether GM crops are adopted or rejected in various parts of the world, because wider public scrutiny and the newness of the science have led to concerns about environmental and health risks of GM crops that must be dealt with at the policy level.

The battle between U.S. and European Union regulations, which feature very different stances on the acceptance of GM crops in food and feed, has been the major determinant of this outcome. In addition, a variety of nongovernmental organizations (NGOs) that are concerned about the influence of multinational corporations, environmental degradation, crop diversification, food safety, globalization, and the influence of U.S. interests are prominent and influential in both the industrialized and developing countries. These NGOs were not nearly as influential during the Green Revolution.

Lessons from the Green Revolution

What can we determine about the prospects for the Gene Revolution by studying the Green Revolution's successes and failures? The Gene Revolution thus far resembles the Green Revolution in the following ways: (1) It employs new science and technology to create crop seeds that can significantly outperform the types of seeds that preceded it; (2) the impact of the new seed technologies can be critically important to developing world agriculture; and (3) for a variety of reasons, these technologies have not yet reached the parts of the world where they could be most beneficial. On the other hand, the Gene Revolution is *unlike* the Green Revolution in the following ways: (1) The science and technology required to create GM crop seeds are far more complicated than the science and technology used to create Green Revolution agricultural advancements; (2) GM seeds are created largely through private enterprises rather than through public-sector efforts; and (3) the political climate in which agricultural science can influence the world by introducing innovations has changed dramatically since the Green Revolution.

The similarities and differences between the Green and Gene Revolutions lead us to speculate that for the GM crop movement to have the sort of impact that would constitute an agricultural revolution, the following goals still need to be met and the related challenges overcome.

1. Agricultural biotechnology must be tailored toward, and made affordable to, developing-world farmers. Unless these conditions are met, farmers may not see that it is in their best interest to use GM crops at all despite the unique benefits those crops could provide.

2. There is a need for larger investments in research in the public sector. Numerous studies have shown the importance of public-sector research and development to aiding agricultural advancements, including the Green Revolution. Partnerships between the public and private sectors can result in more efficient production of GM crops that are useful to the developing world and can expand the accessibility of those crops and their associated technologies to developing-world farmers.

3. To garner the level of public interest that can sustain an agricultural revolution, agricultural development must once again be regarded as being critically important from a policy perspective in both donor and recipient nations. As population numbers continue to increase today, agricultural development is more necessary than ever to eliminate malnutrition and prevent famine, particularly in sub-Saharan Africa. GM crops are seen by many as a means for addressing those problems. However, policymakers worldwide are far from being a combined force on this issue.

4. Policymakers in the developing world must set regulatory standards that take into consideration the risks as well as the benefits of foods derived from GM crops. This goal is crucial to the cooperation of the many stakeholders that are affected by GM crops and also for the sustainability of the GM crop movement in the foreseeable future. Without regulations that explicitly take into account potential benefits to both farmers and consumers, those nations that might stand to benefit most from GM crops may be discouraged from allowing them to be planted.

Revised regulations on genetically modified crops must accompany widespread collective policy efforts to revitalize agricultural development. And before developing world farmers and consumers can benefit from GM crops or any other type of enhanced crop breeding, the technologies must be affordable and farmers must understand how to use them.

The GM crop movement must overcome an intertwined collection of challenges before it can have an impact beyond those regions of the world that already produce excesses of food. If the GM crop movement can overcome these challenges, while proving itself to be acceptably free of adverse health and environmental impacts, it has the potential to provide benefits to farmers and consumers around the globe in previously inconceivable ways, while mitigating the need to use potentially harmful chemicals or scarce water supplies for agriculture. It can then indeed become a true "Gene Revolution."

Acknowledgments

The generous support of RAND and particularly the encouragement of Brent Bradley, James Thomson, Stephen Rattien, and Debra Knopman have made this report possible.

In preparing this report, the authors have benefited from numerous discussions with colleagues within RAND and with representatives from government, academia, industry, and a variety of other institutions. In particular, we thank:

- Susan Bohandy, Research Communicator, RAND
- Elizabeth Casman, Research Engineer, Carnegie Mellon University
- Nancy DelFavero, Research Editor, RAND
- Anita Duncan, Administrative Assistant, RAND
- R. Scott Farrow, Chief Economist, U.S. General Accounting Office
- Daniel A. Goldstein, Product Coordinator, Monsanto Company
- Lowell S. Hardin, Professor Emeritus of Agricultural Economics, Purdue University
- Robert Klitgaard, Dean, RAND Graduate School, and Professor of International Development and Security
- J. David Miller, Professor of Biochemistry, Carleton University, Ottawa, and Visiting Scientist, Health Canada
- Benoit Morel, Senior Lecturer, Carnegie Mellon University
- M. Granger Morgan, Department Head, Engineering and Public Policy, Carnegie Mellon University

- Robert L. Paarlberg, Professor of Political Science, Wellesley College
- Tina Rapacchietta, Administrative Assistant, RAND
- Anny Wong, Associate Political Scientist, RAND
- Christopher Wozniak, National Program Leader for Food Biotechnology and Microbiology, U.S. Department of Agriculture.

Any errors of fact or judgment are those of the authors.

Acronyms

AATF	African Agricultural Technology Foundation
APHIS	Animal and Plant Health Inspection Service
Bt	*Bacillus thuringiensis*
CGIAR	Consultative Group on International Agricultural Research
CIAT	Centro Internacional de Agricultura Tropical (International Center for Tropical Agriculture)
CIMMYT	Centro Internacional de Mejoramiento de Maiz y Trigo (International Maize and Wheat Improvement Center)
DG	Directorate General
EC	European Commission
EEC	European Economic Community
EPA	U.S. Environmental Protection Agency
ERS	Economic Research Service
EU	European Union
FAO	Food and Agricultural Organization of the United Nations
FAOSTAT	FAO statistics
FDA	U.S. Food and Drug Administration

FFDCA	Federal Food, Drug and Cosmetic Act
FIFRA	Federal Insecticide, Fungicide, and Rodenticide Act
FPPA	Federal Plant Pest Act
FPQA	Federal Plant Quarantine Act
GM	genetically modified
GMO	genetically modified organism
GRAS	Generally Recognized As Safe
Ha	hectare
HYV	high-yield variety
IARC	international agricultural research center
IITA	International Institute of Tropical Agriculture
IP	intellectual property
IRRI	International Rice Research Institute
ISAAA	International Service for the Acquisition of Agri-biotech Applications
Kg	kilogram
MAP	Mexican Agricultural Program
NARS	national agricultural research systems
NAS	National Academy of Sciences
NEPA	National Environmental Policy Act
NGO	nongovernmental organization
OSS	Office of Special Studies
OSTP	Office of Science and Technology Policy
PPA	Plant Protection Act
R&D	research and development
UK	United Kingdom

UNDP	United Nations Development Programme
UNICEF	United Nations Children's Fund
USAID	United States Agency for International Development
U.S.C.	U.S. Code
USDA	U.S. Department of Agriculture
WTO	World Trade Organization

CHAPTER ONE

Introduction

Agriculture is a very old form of human technology. By harnessing sunlight, soil nutrients, and water toward satisfying their wants and needs, human beings for much of their history have made more productive use of agriculture than they ever could have derived from hunting and gathering. Over the millennia, the interaction of agriculture with population growth and dispersion has been at the core of human cultural and economic progress.[1]

For as long as ten thousand years, humans have been purposefully choosing the genetic makeup of the crops they grow. Genetic selection for features such as faster growth, larger seeds, or sweeter fruits has dramatically changed domesticated plant species compared with their wild relatives. Indeed, many of our modern crops were developed before modern scientific understanding of plant breeding (Byrne et al., 2004).

Despite such agricultural improvements, concerns have arisen many times and in numerous places that the population would grow faster than the food available to feed it. Periodic famines supported these fears. In the late 1700s, English economist Thomas Malthus (1766–1834) predicted that population growth, left unchecked,

---

[1] Anthropologists, economists, geographers, and others have long speculated about the relationship between agricultural development and population growth. British economist David Ricardo (1772–1823), for example, theorized that technical improvements in agriculture enabled population growth, whereas Danish economist Esther Boserup argued, alternatively, that population pressure on the land was a precondition for the emergence and development of agriculture (Boserup, 1965).

would lead to famine in human civilizations as a matter of course because the food supply, which is limited by the availability and quality of land, will grow more slowly than the population. Malthusian predictions have not come to pass, partly because of the emergence of improved agricultural technologies, improvements that were scattered and infrequently used before Malthus's time. Indeed, a cultivator from ancient Egypt might well have stepped into a hired hand's role on an American farm as late as the 1880s with only several hours of instruction. Human labor, animal power, and simple implements were still, along with the land, the means of food production.

The Agricultural Revolutions of the 19th and 20th Centuries

In the late 19th to early 20th century, a series of unprecedented technological revolutions transformed agriculture, first in industrialized countries and then more broadly worldwide, although not universally. The grain reaper and cotton gin, and later the tractor and thresher, pushed the *mechanical revolution* of the 1890s forward, increasing the amount of seed that could be planted and the amount of land that could be usefully farmed with the same amount of labor. Then, shortly after the turn of the 19th century, the Haber-Bosch process made possible the economical production of nitrogen fertilizer, whose spreading application in the United States and Western Europe introduced a *chemical revolution* that further increased the yield a farmer could produce with the same amount of seed and land. The first half of the 20th century brought a third set of sweeping changes. Hybrid crop breeding, first done with corn in the United States, created new strains that with increased application of chemical fertilizers substantially boosted production per acre. This *hybrid revolution* eventually extended to many other crops and many other countries.

These three agricultural revolutions arose from technological innovations in industrialized countries and primarily affected the farmers and consumers in those countries. The second half of the 20th century produced a different kind of agricultural transformation, one concentrated in less-developed countries with traditional agriculture. This so-called *Green Revolution* (discussed further in Chapter Two) brought the rapid spread of hybrid wheat and rice, then other hybrid crops, first in Mexico and then in various Asian nations. Production per hectare dramatically increased when the crop was appropriately fertilized and irrigated.

Although they varied substantially in form and scope, these four 19th and 20th century agricultural revolutions shared the following five characteristics:

1. The movements gave farmers incentives to produce; i.e., the technologies provided a net benefit to farmers.
2. The movements substantially improved agricultural production, food nutrition, or both; and/or they substantially decreased necessary resources such as human labor, fertilizer, or water.
3. Farmers were generally willing to adapt culturally and economically to the new technologies, and consumers accepted the products of the new technologies.
4. There was cooperation among those that provided the technologies, regulated the technologies, and used the technologies, and there was support at the governmental level.
5. The movements were sustainable, eventually without public subsidization, and were not just acceptable but were desirable to most stakeholders (i.e., growers, consumers, and the government).

All of these revolutions have by now run a long-enough course to reveal their consequences, both planned and unintentional, and beneficial and harmful. Among the benefits have been substantial increases in food security[2] and rural living standards in much of the world, and the production of agricultural resources used in producing non-agricultural goods and services that are an integral part of mod-

---

[2] Food security refers to both having enough food on the whole and making sure that distribution systems are in place such that the food actually gets to the people who need it.

ern life. But in the process of revolutionizing agriculture, the lives of many millions of people who left the farms for other employment were disrupted, and the advancements were not always to their advantage.

The "Gene Revolution"

It is in the context of the hundred-year history of technological change that we consider the most recent movement in world agriculture: genetically modified (GM) crops, produced through modern biotechnology that enables genes to be transferred across different species and even across different plant kingdoms, to introduce desired traits into a host plant.[3] After just a decade, the GM crop movement is already beginning to revolutionize agriculture in new ways, with previously unachievable benefits and novel potential risks.

This study focuses on the genetically modified crop movement and whether it has the potential to revolutionize agriculture in the developing world and to truly become the "Gene Revolution" that some of its proponents already call it. We focus on the developing world because it is in greatest need of a new agricultural revolution—whether in the form of GM crops or another revolution altogether—given the rapidly growing populations, lagging agricultural technologies, and malnutrition in the world's poorest nations.

Three presumptions motivated this study: (1) Reducing hunger and malnutrition is desirable; (2) now, as in the past, revolutionary technological change in world agriculture can substantially reduce hunger and malnutrition; and (3) now, as in the past, agricultural technologies can be designed and used such that the majority of farmers, consumers, and experts will agree that the technologies are worth their attendant risks.

After a running start in the United States, the progress of the GM crop movement has slowed, and perhaps even stalled, on the

---

[3] GM crops contain genes that are artificially inserted instead of the plants' acquiring them through sexual means. See Chapter Three for a further discussion.

global stage. As just one example, China, the world's first commercial producer of GM crops, has become far more precautionary regarding GM food trade and production. Why is this happening in many places around the world? The major reasons seem clear enough: Many consumers do not want to consume genetically modified agricultural products, some farmers do not want to grow GM crops, influential interest groups advocate against GM crop production and trade, and a number of governments are also against GM crop production. Concerns over ecological and health risks, and the attendant economic risks, explain many of these attitudes. Yet, the same concerns arose in varying degrees about the previous agricultural revolutions, particularly the Green Revolution of a generation ago. It is that particular experience—the Green Revolution—we examine in this study for answers to whether and how the GM crop movement may revolutionize world agriculture.

The Gene Revolution in Light of the Earlier Green Revolution

We believe that the Green Revolution is similar enough to the GM crop movement in terms of purpose, scope, and influencing factors to provide important insights into the future of GM technology. For example, the Green Revolution achieved previously unattainable increases in food production, with important implications for parts of the developing world where food supply was short. The GM crop movement has comparable potential. Green Revolution scientists genetically enhanced existing crops in novel ways that created controversy at their inception, as are the methods of the scientists spearheading the GM crop movement today. The Green Revolution required financial and political support from a variety of stakeholders and decisionmakers, just as the GM crop movement does today.

For this study, we conducted a systematic investigation of the Green Revolution, identifying factors associated with its successes and failures. From this investigation, we identified lessons that can be applied to the current Gene Revolution to provide guidelines on how policymakers, industry leaders, and other key decisionmakers can minimize the risks and maximize the benefits from this agricultural revolution. In short, we explore the question: What can be learned about whether and how the Gene Revolution can succeed on a global scale by studying the successes and failures of the Green Revolution?

Our analysis of the GM crop movement's Gene Revolution and its Green Revolution predecessor is structured around four main areas of comparison:

- The science and technology of each movement
- Their sources of funding and financial investment
- Where each agricultural movement took place
- The political environment surrounding these movements.

Science and Technology

Long before scientists and engineers turned their research tools to agriculture, farmers worldwide had already developed and adapted yield-maximizing techniques, such as fighting weeds, spreading manure, rotating crops, leaving land fallow for a period of time, and setting aside seeds from the sturdiest plants to sow the following season (Pinstrup-Andersen and Schioler, 2001). It was advances in science and technology, however, which allowed for the recent agricultural revolutions to occur. Mechanical, chemical, plant breeding, and now genetic sciences have enabled agricultural transformations that have greatly increased yield and reduced labor requirements.

Funding and Sources of Financial Investment

The type of financial support for agricultural research and development has a major impact on how new technologies are created and disseminated. The source of research and development funding matters greatly because it influences public attitudes, governmental willingness to adopt new technologies, and the types of technologies that are developed (which may be useful or useless in certain parts of the world, depending on the technology).

Some key questions in this area are: Who is providing the funding for the new science and technology? Is the capital investment for profit or for philanthropy? How are the funding institutions organized—do they work together or separately to achieve their aims? How can the industrial and developing world organize for the purpose of funding agricultural technologies? The answers to these questions will determine whether the Gene Revolution will have the sustained financial and political support needed to transfer the technology worldwide.

Where the Revolution Takes Place

Many factors influence where an agricultural revolution takes place. First, the scientific and technological developments may have limited geographic applicability. For example, a pesticide that provides protection against a specific pest is useful only in areas where that pest does significant damage. Likewise, soybeans that are genetically modified to tolerate a specific herbicide are useful only where soybeans are planted and where the particular herbicide is commonly used. Indeed, because crops are planted in such a wide variety of agronomic conditions, producers usually are unable to develop agricultural technologies that are beneficial on a global basis. Aside from scientific considerations, the agricultural, trade, and consumer policies that weaken farmers' production incentives can make even the most usable technology unprofitable and therefore unusable. Low farming income, the high cost of complementary inputs (e.g., pesticides, fertilizers, and irrigation), and badly defined land property rights can also prevent adoption of new technologies.

Whether the people of a particular region are willing to accept a new technology and adapt their lifestyles accordingly is a significant factor in whether the technology makes any headway. Certain local farming practices have become tradition for good reason and are not easily altered. In some cases, those practices shape a community's value system. Advancing agricultural technologies can, for example, disrupt daily and seasonal routines of work and leisure, particularly the division of work among men, women, and children in a household. Also, some stakeholder groups will benefit and others will suffer from widespread technological change.

Land ownership issues are crucial in this area. For example, the mechanical revolution of the 19th and 20th centuries replaced human labor in the fields with machines, first in the industrial world and then in the developing world. Only those farmers who could afford the machines and who had access to enough land to make using the machines worthwhile were able to thrive; smaller farmers and those who were unwilling to adjust to technological change often sold or lost their land. That trend may hold for future agricultural movements as well.

Policies and Politics

Food supply and food security have always figured prominently in the strength and stability of a nation-state (Perkins, 1997). Internal stability in peacetime is heavily dependent on a safe and steady food supply, and the advent of war brings the continued dependability of the food supply into sharp focus. Policies and regulations can help to either mobilize an agricultural revolution or stymie it.

The interrelationship of the many different governing bodies is even more important today than it was a generation ago; food regulations in Europe, for example, did not influence whether Mexico or China adopted Green Revolution technologies. Today, with the conflicts among various governing bodies regarding the safety or desirability of particular agricultural technologies (such as GM crops) and their food products, adoption of those technologies may slow worldwide or stop completely. When there is harmony among policymakers about the desirability of promoting an agricultural movement, that movement stands a stronger chance of making a revolutionary impact.

ORGANIZATION OF THIS REPORT

Chapter Two describes the Green Revolution along the lines of the topics discussed above: science and technology, funding, where the revolution took place, and policies and politics surrounding the revolution. Chapter Three describes the ongoing Gene Revolution along the same lines, with the understanding that this revolution is ongoing and thus still evolving. Chapter Four applies the lessons from Chapter Two to the Gene Revolution and derives implications for policy in the areas of science, agriculture, regulation, and technical assistance, in both developing and developed countries.

The Green Revolution

Green Revolution is the term used to describe the spread of new agricultural technologies that dramatically increased food production in the developing world beginning in the middle of the 20th century, the impact of which is still felt today. In the 1940s, to address the problem of impending famine from a growing imbalance between population and food supply, the Rockefeller Foundation provided funds for agricultural advances in developing nations and gathered together a team of dedicated researchers from various parts of the world. Other national and international institutions joined in this effort over the following decades.

The crops that were developed as part of this effort were superior to other locally planted crops in yield increase, yield stability, wide-scale adaptability, short growing season duration, resistance to biotic stresses (diseases and insects), tolerance to abiotic stresses (drought and flooding), and grain quality (Khush, 2001). Importantly, fertilizers, pesticides, and irrigation systems were also introduced in many parts of the developing world, and this whole technology package was supported in many parts of Asia and Latin America by subsidies and government-guaranteed product prices.

The Green Revolution has been credited with "making an astounding jump in basic food crops in the developing world, especially Asia" (Baum, 1986, p. 1). It significantly increased food production in Asia and Latin America at a time when massive malnutrition in those areas was feared. Its success was due to a combination of circumstances that made the time ripe for an agricultural revolution to take place in much of the developing world. Worldwide, policymakers saw increased food production as a priority after World War II and set the wheels in motion to facilitate that goal. At the same time, agricultural science and technology in the industrial world—which produced hybrid seeds, pesticides, fertilizers, and irrigation systems— had advanced to the point that new technologies could be transferred to other parts of the world where they could be useful. Indeed, universities in the United States played a crucial role in the advancement of agricultural development in that they were a source of scientists and a provider of education, particularly advanced degrees.

This chapter describes the science and technology, funding sources, worldwide spread, and policies and political motivations of the Green Revolution. We describe the factors in each of these areas that contributed to increased food production and explain how those factors were intertwined. At the end of the chapter, we provide lessons from the Green Revolution for future agricultural revolutions (including the ongoing Gene Revolution) and describe the continual problems that the Green Revolution was not able to solve—leaving challenges for today's agricultural scientists and policymakers.

Science and Technology

Two of the criteria for an agricultural revolution are a substantial improvement in food production and sustainable use of the relevant technologies. The Green Revolution, in the many places that it was adopted, met these criteria very well. What made the Green Revolution technologies successful in bringing about revolutionary change was the integration of three factors: (1) plant breeding methodologies that were adapted in innovative ways to produce regionally useful crops, (2) scientists and researchers who combined a range of various technologies to achieve their objectives, and (3) local scientists who were given the necessary training to breed the crops themselves. These three factors, along with programs and policies that kept farmers informed and protected their production incentives, were responsible for a significant increase in food production.

Plant Breeding Methodologies

The plant breeding research that led to high-yield varieties (HYVs) of staple food crops was perhaps the most important factor in the food production increase. Several genetic traits were selected to increase yield, yield stability, and wide-scale adaptability of rice, wheat, and maize.

Three types of breeding strategies were used to develop these three types of crops. The first was conventional breeding, in which parents of the same line were crossed and the resulting offspring were screened for desirable traits, and those offspring that contained the traits for the next round of crossing were selected. The second strategy was hybrid breeding, in which plants from different lines were crossbred to produce offspring with increased vigor. The third strategy, developed in Mexico, was the innovative technique of *shuttle breeding*, which at the time violated the conventionally held belief that plants had to be adapted to the native soils and climate in order to be successful, and only if all selections were from plants that grow in the actual area where the crossbred crops are to be produced could the new varieties be well-adapted to their environment (Perkins, 1997). Shuttle breeding, in contrast, involves growing plants at various locations and crossbreeding between those lines to obtain offspring with widely adaptable traits that can be grown under diverse growing conditions, and, more specifically, with improved resistance to the devastating disease of wheat rust (Khush, 2001). As a result, new high-yielding varieties of wheat, maize, and rice were developed around the world with a better ratio of grain to straw, shorter and sturdier stems, and better response to fertilizers (Pinstrup-Andersen and Schioler, 2001). Notably, Norman Borlaug, the young plant pathologist and plant breeder who developed the shuttle breeding technique, went on to win the Nobel Peace Prize for his Green Revolution work.

Combined Technologies

The combination of irrigation systems, fertilizers, and pesticides, along with the HYV seeds, was also key to the increase in food production (Pingali and Heisey, 2001). When the HYVs were grown with these other inputs, they produced substantially higher yields (Cleaver, 1972; Conway, 1998). Aside from increased yield in most situations, another benefit that the combined Green Revolution technologies provided was greater yield consistency in the face of differing agricultural conditions. For example, the Centro Internacional de Mejoramiento de Maiz y Trigo (CIMMYT) developed a maize variety that gave the second-best yield in Mexico when external planting conditions were optimal but inputs were less than optimal. However, even in times of drought, the CIMMYT variety, combined with the appropriate chemical inputs, provided by far the best yield (Pinstrup-Andersen and Schioler, 2001).

Table 2.1 illustrates the changes in adoption rates of various Green Revolution technologies—acreage and percentage of new high-yield crop varieties, irrigation, fertilizer consumption, number of tractors, and cereal production—in developing countries in Asia.

Table 2.1
Changes in Production Inputs in India, Pakistan, and China Due to the Green Revolution

	Adoption of High-Yield Wheat Varieties (million ha/% of total growing area)	Adoption of High-Yield Rice Varieties (million ha/% of total growing area)	Irrigation (millions of ha)	Fertilizer/ Nutrient Consumption (millions of tonnes)	Tractors (millions)	Cereal Production (millions of tonnes)
1960	0/0	0/0	87	2	0.2	309
1970	14/20	15/20	106	10	0.5	463
1980	39/49	55/43	129	29	2.0	618
1990	60/70	85/65	158	54	3.4	858
2000	70/84	100/74	175	70	4.8	962

SOURCE: Borlaug, 2003.
NOTES: ha = hectare (equivalent to 2.47 acres); tonne = metric ton.

Training of Local Scientists

Local scientists were given the necessary training to breed the crops themselves so that they would be able to advance the agricultural revolution in their own nations without continued supervision from the outside. The Food and Agricultural Organization of the United Nations (FAO) and the Rockefeller Foundation established a training program that recruited young scientists from around the world to learn the relevant agricultural disciplines. After their training was completed, the scientists were given HYV semi-dwarf seeds to take back to their respective nations (Perkins, 1997; Borlaug, 2003).

As a result of these successful new technologies and their transfer, food production improved enormously over the course of only a few decades in many parts of the world. Taking Latin America as an example, based on the combination of HYV wheat varieties, chemical inputs, and expanded acreage, total maize production increased by 142 percent and total wheat production increased by 100 percent between 1961 and 1991 (according to FAO statistics [FAOSTAT]). Even earlier, Mexico was able to achieve self-sufficiency in wheat and maize production by 1956 through Green Revolution technologies (Cleaver, 1972).

In Asia, HYV wheat and rice were first tested in the early 1960s, and production campaigns were launched soon afterward. These campaigns led to a sharp increase in adoption of HYV wheat and rice, yields, and production. In South Asia, for example, rice production more than doubled from 1961 to 1991, and wheat production almost quadrupled in that time (according to FAOSTAT).

Funding

How the Green Revolution was financed can be seen as a twofold process: funding the research to develop the technologies and to transfer them to various parts of the world, and then financing the agricultural technologies in those regions so that the local farmers could afford to use them. International and domestic funders (i.e., funders within nations where Green Revolution technologies were being promoted) were key to this process. International funders acted first to provide the necessary monetary, scientific, and human capital, and domestic funders cooperated with these international groups to sustain the agricultural movement within particular regions. Importantly, the international sources of finance were public organizations, both governmental and nongovernmental, acting ostensibly out of public interest rather than from business-related motives.

Funders and scientists being involved in helping other nations was a rather new concept at the time. Prior to World War II, there was little support for nations working together to provide finances for worldwide industrialization and agricultural modernization. An important exception was the Rockefeller Foundation, which already had a program in China in the 1920s to improve wheat production (Perkins, 1997). But the outbreak of World War II changed the philanthropic programs of the Rockefeller Foundation as well as those of other funders; as such, their efforts to improve agriculture took a global turn. Public-sector funding and leadership turned out to be crucial to ensuring the success of the Green Revolution. The Rockefeller Foundation in the 1940s is generally regarded as the "founder" of the Green Revolution, with its programs committed to agriculture in the developing world (Pinstrup-Andersen and Schioler, 2001). Later, the World Bank, the FAO, and the United States Agency for International Development (USAID) joined in the effort (Cleaver, 1972).

At the country level, success was usually due to collaboration between the international agricultural research centers (IARCs), governments, and their national agricultural research systems (NARS). The IARCs gathered test nurseries of selected seeds, distributed them widely to the NARS, and analyzed and shared the results of the seed development. NARS, thus, had in their hands a wide selection of germplasm for use in their own breeding programs from which to develop locally adapted HYVs (Hardin, 2003).

Indeed, an important factor that contributed to the sustained success of the Green Revolution was that scientists received financial support from these public institutions to set up research centers in developing nations. As such, they had firsthand information on what the local farmers needed and what the local planting conditions were. Furthermore, they were on site to train local farmers and scientists to ensure that knowledge of how to use these new technologies would be passed on.

Recognizing the importance of developing local expertise, organizations in developed countries pooled their resources to aid in the effort to link industrial-world scientists with developing-world locations, which resulted in the formation of the Consultative Group on International Agricultural Research (CGIAR). CGIAR now includes a system of 16 Future Harvest Centers working in more than 100 countries conducting research on all key crops of the developing world and on livestock, fish, forestry, plant genetics, and food policy (Pinstrup-Andersen and Schioler, 2001). With the creation of CGIAR, support for the Green Revolution became more broad-based and included European nations, Canada, and Japan. Now some 50 donors, including the World Bank, the United Nations, the United States Agency for International Development, and other public organizations, have provided expanded funding for the Green Revolution through and in association with CGIAR.

Farmers in Latin America and Asia were able to afford the relatively expensive HYV seeds, pesticides, and fertilizers through a combination of internationally provided subsidies and low-interest loan systems. Adoption of the Green Revolution package was relatively expensive; in Bangladesh, for example, the input costs were 60 percent more than those for traditional seed varieties. Small subsistence farmers could afford these packages only if they borrowed money, which usually came at a high interest rate. Establishment of rural banks was important for adoption of the Green Revolution technologies in Asia and Central America (Conway, 1998).

The substantial production improvements that farmers experienced through these Green Revolution technologies enabled them to pay off their loans quickly and easily. Moreover, the money earned from the increased food production often meant that whole villages could make investments in their schools and roads (Kilman and Thurow, 2002), efforts that in the long run would further improve not only agricultural output but also many other aspects of a community's socioeconomic well-being.

Where the Green Revolution Occurred

The Green Revolution is generally considered to have been a success in four distinct regions of the world: Latin America, China and Southeast Asia, India and South Asia, and the United Kingdom (UK). Except for the UK, success was generally defined by a production increase that staved off potential malnutrition, quite apart from concerns about the environment, socioeconomic equality, or other such issues. Millions of lives were at risk from malnutrition in Asia and Latin America immediately after World War II, and results were needed, fast. An increase in food production had to be the top Green Revolution priority in those areas (Pinstrup-Andersen and Schioler, 2001).

Latin America

Latin America was a primary region of interest during the Green Revolution because of a combination of need and local governmental interest in improving agriculture. The European conquest and its centuries-long aftermath had resulted in depletion of soil nutrients and other natural resources that were necessary for sustainable crop yields. The Rockefeller Foundation created the Mexican Agricultural Program (MAP) in 1943 amid concerns about political stability and national security as well as the specter of malnutrition.

The Ford and Rockefeller Foundations and the Mexican government signed a memorandum of understanding in Mexico in early 1943 to establish the Office of Special Studies (OSS), a follow-up to MAP, as a semi-autonomous research unit within the Mexican Department of Agriculture. Until its transformation in 1966 to the CIMMYT, the OSS was the research center that spurred a major transformation of Mexican agriculture (Perkins, 1997), increasing yields in wheat and maize to an extent that effectively prevented widespread famine.

Figure 2.1 shows the change in maize and wheat production in Latin America and the Caribbean from 1961 to 1991. The primary improvements in Mexico came before 1961; the rest of Latin America and the Caribbean soon followed suit. Over those 30 years, production of both wheat and maize approximately doubled. In more recent decades, production increases in maize have leveled off; however, wheat production has continued to increase steadily. During this same 30-year period, the combined populations in Latin America and the Caribbean increased from about 417 million in 1960 to about 718 million in 1990 (Brea, 2003)—a 72 percent increase that is more than matched by the increase in wheat and maize production.

Asia

The Green Revolution also achieved significant agricultural yield increases throughout Asia. Before the new agricultural technologies and

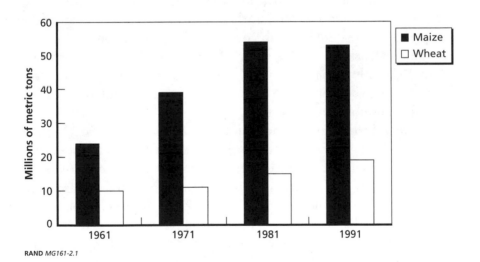

RAND *MG161-2.1*

Source: FAOSTAT.

Figure 2.1—Total Production of Maize and Wheat in Latin America and the Caribbean, 1961 to 1991

practices were brought to the continent, wheat yield in India and China was on par with that of Europe during the Middle Ages (600–800 kilograms per hectare [kg/ha]). This yield was adequate prior to 1960, because the population had stayed within the bounds necessary to ensure adequate food supplies even with suboptimal agricultural practices. In the 1960s, however, India and China were experiencing improved living conditions and modern health-care techniques that prolonged life expectancy, which, compounded with high fertility, could have led to massive malnutrition (Pinstrup-Andersen and Schioler, 2001).

Fortunately, at that time, rice production increased substantially throughout much of East and Southeast Asia, largely from strains developed at the International Rice Research Institute (IRRI) in the Philippines. Founded in 1960 by the Ford and Rockefeller Foundations, IRRI began its research activities in 1962 with the development of semi-dwarf breeding lines for rice. (Semi-dwarf strains have shorter, sturdier stems to prevent them from bending and lodging the rice grains in the ground.) Subsequently, rice production throughout East and Southeast Asia grew from 52 million metric tons in 1961 to about 125 million metric tons in 1991 (as shown in Figure 2.2)—a much faster growth rate than that of the regional population, which had almost doubled over that time.[1]

When India gained independence from Britain in the late 1940s, Pakistan was partitioned into an independent nation, causing India to lose major areas of irrigated wheat land in the west, vast rice-producing areas in the east, and important agricultural research and education facilities. The country was facing a crisis. N. C. Mehta, secretary of the Indian Council of Agricultural Research, stated in 1951, "The country can rise as a whole only if our agricultural economy with its millions of farms and *lakhs* of villages is to revive with a new sense of energy and well-being" (Perkins, 1997).

---

[1] International Rice Research Institute website (http://www.irri.org/about/impact.asp).

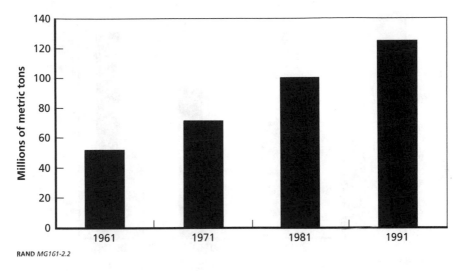

RAND *MG161-2.2*

Source: FAOSTAT.

Figure 2.2—Total Rice Production in East and Southeast Asia, 1961 to 1991

In the two decades that followed the partitioning, India tried several methods to deal with potential food shortages: expanding the amount of farmland, establishing price controls, launching community development programs, and obtaining grants and low-cost sales of surplus crops from the United States and other nations. Unfortunately, other government policies—import substitution, cheap food, and industrialization at the expense of agriculture—rendered these efforts largely ineffective (Cleaver, 1972; Farmer, 1986). After some initial hesitancy, parts of India and other South Asian nations finally embraced Green Revolution technologies.

Those measures paid off. Figure 2.3 shows the impact of the Green Revolution on rice and wheat production in South Asia. Rice production more than doubled from 1961 to 1991, and wheat production almost quadrupled in that time, more than matching the approximate doubling of the population in South Asia.[2]

---

[2] International Rice Research Institute website.

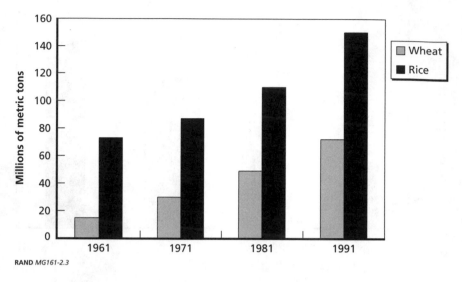

RAND *MG161-2.3*

Source: FAOSTAT.

Figure 2.3—Total Wheat and Rice Production in South Asia, 1961 to 1991

Thus, steady and significant increases in crop production throughout East, Southeast, and South Asia, as in Latin America, prevented the malnutrition that might have otherwise occurred with the soaring populations of the second half of the 20th century.

The United Kingdom

The United Kingdom provides an example of the Green Revolution's significant impact on the industrial world. In the post-Colonial era from 1935 to 1954, Britain was concerned over whether its lands would produce sufficient food for its population because it had previously depended on imports from the colonies. It was especially concerned about wartime food production in the early 1940s (Perkins, 1997). Hence, throughout the United Kingdom, growers adopted the Green Revolution technologies, particularly the high-yield crop varieties. The adoption of these varieties paid off: Britain produced only 23 percent of its wheat consumption in 1936–1939, but had boosted that rate to 67 percent in 1974–1975 and to 77 percent in 1980–1981 (Perkins, 1997). Whether or not an actual food crisis would have occurred without the Green Revolution technology, Britain did become more independent with regard to its key food crops.

A Failure in Africa?

Despite its successes in many parts of the world, the Green Revolution has not yet made a significant impact on most areas of Africa. In the later part of the Green Revolution, four CGIAR centers were founded in Africa. These centers have contributed to increases in the production of maize, bananas, cassava, and rice. But these increases have been limited, and the growing population has effectively neutralized these gains (Pinstrup-Andersen and Schioler, 2001). Overall, while average agricultural production in Asia after the Green Revolution has increased dramatically to nearly three tonnes per hectare from 1960 to 1990, Africa's production level has fallen to about one tonne per hectare, which is about the average productivity of British farmers during the reign of the Roman Empire (Conway, 2003).

Why hasn't the Green Revolution been more effective in Africa? Research that successfully identifies the constraints that had an actual effect on the African situation has not yet been reported. However, we postulate several factors that might explain this lack of success in terms of the criteria for agricultural revolutions.

One possibility is that sub-Saharan Africa features types of agriculture that could not benefit from the technological packages developed through the Green Revolution for use in Asia and Latin America. Therefore, the criterion of substantial production increases could not be met. The new HYV crops that were developed elsewhere in the world were not suited to African planting conditions, where the topsoil is thinner and weather patterns, such as periods of drought, are more unpredictable (Pinstrup-Andersen and Schioler, 2001). Serious and uniquely African agricultural problems persist. The staple crop—maize—is attacked by insects, streak virus, and the parasitic weed Striga that extracts nutrients from roots. Mealy bugs and mosaic virus attack cassava. Weevils, nematodes, and the fungal disease black Sigatoka attack bananas. Fungal diseases shrivel and weaken bean pods, and drought is regularly occurring (Pinstrup-Andersen and Schioler, 2001). These problems would have been difficult to overcome through even the enhanced conventional means that the Green Revolution employed. A number of the CGIAR centers, such as the International Institute of Tropical Agriculture (IITA) and the Centro Internacional de Agricultura Tropical, or International Center for Tropical Agriculture (CIAT), have developed improved varieties of sorghum, millet, cowpeas, cassava, potatoes, and sweet potatoes for African use. But the sorts of dramatic yield improvements scored with rice and wheat in Asia have not been achieved in Africa.

Also, infrastructural problems have continued to serve as a barrier to Green Revolution success in Africa. Inadequate property rights protection have had the effect of blunting farmers' production opportunities and incentives, leaving little possibility for private gain in adopting new technologies. Local banks have been unable or unwilling to assist in providing the necessary loans for farmers to purchase new technologies. It is also possible that cultural constraints that control particular divisions of labor among family members or particular patterns and timing of planting, cultivating, and harvesting may be stronger in Africa than in Asia and Latin America.

Even if there were a greater degree of cooperation among the various African institutions in promoting Green Revolution practices, and even if agricultural technologies would have proven to be more successful, geography alone would limit the Green Revolution's success in Africa. In the 1700s, Adam Smith, in his *Wealth of Nations*, wrote that the inland parts of Africa would lag in economic development because of problems with transportation of goods. Moreover, geography may affect other important determinants of agricultural development, such as communications, human and animal health, natural resources, and population density, as well as transportation (Sachs, 1997). The African transportation system was poorly prepared to deliver Green Revolution technologies to the places they were needed. Uganda and Ethiopia, for example, had fewer than 100 kilometers of paved roads per one million people as of 2001—roughly 100 times fewer kilometers of roadway than in other developing nations such as Brazil and India, and some thousand times less than in industrialized nations such as the United States and France (Borlaug, 2003). In addition, many of the roads in Africa were designed to lead to mines rather than farmlands. Because of these transportation obstacles, Green Revolution technologies could not reach African farmers easily or reliably.

Partly as a result of these limitations, malnutrition has persisted and indeed grown even worse in sub-Saharan Africa, unlike in other regions of the world where malnutrition is on the wane. Despite population increases over the past three decades, the percentage of chronically malnourished has declined in most areas of the developing world, except for Africa, as Table 2.2 shows.

Through the efforts of the CGIAR centers that were established in Africa and other important measures, agricultural development eventually began with specific crops that are of particular importance to the diets of sub-Saharan Africa. Table 2.3 indicates that significant

Table 2.2
Change in Percentage of the Chronically Malnourished Population in Developing Countries, 1970 to 1990

	East Asia	South Asia	West Asia/ North Africa	Sub-Saharan Africa	Latin America
1970	44%	34%	24%	35%	19%
1990	16%	24%	8%	37%	13%

Source: Conway, 1998.

Table 2.3
Publicly Funded Agricultural Research, by Country, Crop, and Year Research Began

Country	Research Target	Year Research Began
Chile	Wheat and maize	1940
Mexico	Wheat	1943
Peru	Maize	1954
Brazil	Soybeans	1955
Colombia	Rice	1957
Bangladesh	Wheat and rice	1961
Philippines	Rice	1966
Pakistan	Wheat	1967
Rwanda	Potatoes	1978
Senegal	Cowpeas	1981

SOURCE: Conway, 1998.

investments in African crop improvements did not begin until the late stages of the Green Revolution. The table lists the years in which research began in earnest for certain crops in various countries.

In summary, it may be that the Green Revolution has not *failed* in Africa; rather, it has yet to be delivered to Africa in a way that will truly revolutionize agriculture, for the reasons stated above.

Policies and Politics

It has been argued (Farmer, 1986; Perkins, 1997; U.S. Department of Agriculture [USDA], 2003) that Green Revolution efforts came to fruition in a particular context characterized by fear of famine, over-population, and the threatened rise of communist governments in areas considered a strategic threat to the West. If so, then motivations for the Green Revolution reach far beyond the agricultural and phil-anthropic realm. A number of various political motivations likely shaped the success of the Green Revolution in certain parts of the world, demonstrating on a larger scale the importance of cooperation among the various stakeholders to enable the agricultural revolution to succeed. Governmental agendas, both domestic and international, were altered to accommodate the new agricultural advances.

Domestic Interests
In the mid-20th century, it was unquestioned that greater food production would lead to greater political stability globally as well as to greater prosperity and security in developing nations. Various nations at various socioeconomic levels all saw the need for vastly increased agricultural production, for somewhat different reasons (Farmer, 1986).

In addressing the situation in Mexico, for example, Perkins (1997) argued that beyond the desire to produce more food, the Mexican Agricultural Program's primary proponents within Mexico wished to completely reshape the Mexican economy and intended to use Green Revolution technologies as a means to that end. Namely, to create a modern industrial state, agriculture would need to make use of new technologies that were more labor-efficient so that a large number of farmers could choose (or would be forced) to leave the rural areas and become part of the industrial workforce. Former human labor from agriculture would have to provide at least part of the capital for Mexican industrialization.

India may have embraced the Green Revolution partly to recover from the damages caused by British imperialism and the shattering of the economy of British India after its independence (Farmer, 1986; Perkins, 1997). The security and autonomy of the Indian nation and foreign exchange considerations were the prime drivers in the national commitment to support crop breeding. Although Gandhi had envisioned an India devoted almost entirely to agriculture, many Indian leaders wished to develop the nation's industrial sector for the purpose of remaining competitive in the world market. They decided that if industrialization were to occur, the wealth of the countryside had to finance most of that industrial development—hence, the need for improved food yield. Moreover, the major impetus for the rapid uptake of HYV seeds in South Asia was the two consecutive crop failures during the monsoons of 1966 and 1967. The United States, which was the only country in the world carrying food reserves, shipped much of its surplus grain to India during those years; however, such dependence on foreign aid was recognized by both India and the United States as being risky and undesirable (Conway, 1998).

The UK, too, expanded its commitment to crop breeding as it struggled to reconstruct its post-imperialist economy. After the loss of its significant food-producing colonial lands worldwide, the British government saw the need for an independent food supply as an important component of its political agenda in the mid-20th century. Prior to that time, Britain's only prosperous area of agriculture was livestock production. However, the emergency times of World War II increased the need for British food production of crops, and the loss of India, in particular, as a food producer in the late 1940s further brought about policy pressures to improve agricultural production through Green Revolution methods (Perkins, 1997).

International Interests
It has also been argued (Lipton, 1996; United Nations Development Programme [UNDP], 1994) that much of the development aid worldwide has supported large defense-oriented or commercial projects, with the improvement of living standards as a secondary goal at best. A case in point concerns aid given for the Pergau Dam project in Malaysia. The Overseas Development Administration in the United Kingdom had advised strongly against giving aid to this project because of its low developmental value in the region, but its opinion was overruled on the grounds of safeguarding British defense contracts (Lipton, 1996). Although this is a different situation from the agricultural pursuits in developing nations, similar national security concerns influenced the decision to provide aid for the agricultural technology transfer.

Although the Green Revolution did in fact significantly reduce hunger and poverty in many parts of the developing world, the international motivation for its success extended beyond the goal of poverty reduction. In particular, the United States made significant commitments at the federal and philanthropic levels to promote crop breeding as part of the Cold War effort to contain the spread of communism and possibly to foster other economic and foreign policy objectives. President Harry Truman's 1949 inaugural speech emphasized the need for nations to unite against a communist force: "The United States and other like-minded nations find themselves directly opposed by a regime with contrary aims . . . that false philosophy is communism." To that end, Truman proposed four major courses of action, the fourth of which concerned alleviating hunger and disease in underdeveloped areas: "More than half the people of the world are living in conditions approaching misery. Their food is inadequate. They are victims of disease. Their economic life is primitive and stagnant. Their poverty is a handicap and a threat both to them and to more prosperous areas . . . . Greater [food] production is the key to prosperity and peace" (Truman, 1949).

Communism was not the only political system that Green Revolution visionaries wanted to thwart; other socialist or radical movements were also to be quelled. There seemed to be two theories about the purpose of the Green Revolution in Mexico. One was genuine philanthropy—alleviating the poverty of Mexico's masses and thus empowering them to make choices that would improve their lives. The other was the United States' intent to restructure the Mexican economy toward industrialization. The United States envisioned three ways that the Mexican government could evolve: (1) toward liberal democratic capitalism and an industrial economy, (2) toward a reversion to the quasi-feudal oppression of the *hacendados* (wealthy plantation owners), or (3) continued socialist radicalism of the Cárdenas era (Lázaro Cárdenas was Mexico's president from 1934 to 1940) (Perkins, 1997). Cleaver (1972) suggested, more cynically, that American aid in Mexico was part of the postwar effort to contain socialism specifically to "make the world safe for profits."

In the 1940s, the Rockefeller Foundation also wanted to foster the development of liberal democratic capitalism rather than see either socialism or fascism make further inroads. Although the Ford and Rockefeller Foundations wanted to make it clear to India that they were independent and different from the United States government, in reality they shared the common goal of both humanitarian outreach and "thwarting a perceived threat of communist subversion and keeping India from going the way of China" (Perkins, 1997).

Clearly, agricultural technology transfer in developing nations during the Green Revolution was linked too strongly to political and economic events elsewhere to be analyzed on its own. Political support for improved crop breeding was linked to national security planning and to the need for countries to manage their foreign exchange. Finally, it is interesting to note that Norman Borlaug, the most prominent plant scientist of the Green Revolution and director of the MAP wheat improvement program from 1944 to 1960, was awarded a Nobel *Peace* Prize in 1970 for his agricultural achievements in the developing world. Indeed, the Nobel committee must have considered that food production is linked directly with domestic and international security, and that better agricultural yield in several key nations means a more peaceful world for everyone.

Where the Green Revolution Fell Short: Remaining Challenges

While the Green Revolution had a dramatic and lasting effect for the better, it did not solve agricultural problems worldwide. As discussed above, Africa is a case in point. The Green Revolution also left a number of human health problems unsolved, and, in fact, has exacerbated certain socioeconomic and environmental problems. In the same way that we can learn from all that the Green Revolution did right, we can also draw lessons for the future from the new problems it created and the issues it left unresolved.

Agricultural Challenges

The effectiveness of the Green Revolution is reaching a natural limit, even as human populations in parts of the developing world continue to rise. The population of the industrialized world may increase by only 4 percent from 2000 to 2020, reaching about 1.24 billion. The population of Africa, on the other hand, may well increase by 50 percent in that time and the population of the developing world as a whole by 36 percent. By the year 2030, it is estimated that more than eight billion persons may populate the world (United Nations Population Division, 2002). It is noteworthy that Africa, due to the lagging success of the Green Revolution on that continent and its greatest share of the population increase, is most in need of a new agricultural revolution.

How will agricultural productivity keep up with this rate of population growth? Studies have shown enormous productivity gains in the early decades of the Green Revolution, followed by slower gains more recently. Figure 2.4 shows the changes in annual rates of growth in the yield of major cereals grown in China, one key nation targeted by Green Revolution scientists. One can see that although agricultural growth has continued to be positive in the past decade, the rate of growth has slowed dramatically.

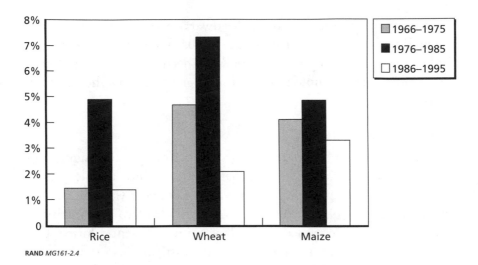

RAND *MG161-2.4*

Source: Pingali and Heisey, 2001.

Figure 2.4—Annual Growth Rates in Yield for Rice, Wheat, and Maize in China, 1966 to 1995

Even though adoption of HYV varieties and overall food production in various parts of the world are clearly increasing, as evidenced by the figures and tables in this chapter, the possibility of a continuing rate of increase that would be necessary to feed ever-growing populations is diminishing (Conway, 2003).

Aside from technological limitations, farmland limitations make future increases in the food supply difficult to achieve. Many agricultural sectors of Africa and Asia have no suitable untilled land left. In 1961, there were 0.44 hectares of farmland for every person in the world. In 2000, the number fell to 0.26 hectares, and projections indicate that by 2050 the number will fall to 0.15 (Pinstrup-Andersen and Schioler, 2001). Thus, from the standpoint of available arable land, the situation does not look promising for the future, if the technological status quo continues as it has. To increase production, it is necessary to improve yields on currently existing fields, beyond what Green Revolution technologies have been able to provide.

Norman Borlaug acknowledged the Green Revolution's limitations in a 2003 speech: "There has been no great increase in the yield capacity of wheat and rice since the dwarf varieties sparked off the Green Revolution of the 1960s and 1970s. In order to meet mankind's rapidly escalating need for food, we need to come up with new and appropriate technological methods for increasing the yield capacity of grains."

Human Health Challenges

Although the Green Revolution did stave off hunger to a significant extent on two continents, an estimated 800 million people still do not have access to sufficient food to meet their needs (National Academy of Sciences [NAS], 2000a), which will lead to malnutrition. Chronically undernourished people have an insufficient caloric intake to meet their daily energy needs, and malnutrition also results from lack of proteins, vitamins, minerals, and other micronutrients in the diet. The most important sources of calories are cereals, the types of crops for which plant breeders developed HYVs during the Green Revolution; cereal diets that supply sufficient calories usually provide enough protein as well. Malnutrition can lead to death, not just through starvation, but because a poor diet leaves humans especially vulnerable to illnesses such as diarrhea, measles, respiratory infections, and malaria (Conway, 1998; Pinstrup-Andersen and Schioler, 2001).

Children whose mothers were malnourished and are themselves lacking basic nutrients and calories are at further risk of a number of developmental disorders. The United Nations Children's Fund (UNICEF) (1998) estimated that malnutrition is the cause of about 12 million deaths annually of children under the age of five in developing countries.

Aside from not having enough to eat in terms of calories, people in developing nations often eat unbalanced diets. For example, those who eat primarily boiled rice or maize porridge rarely consume enough meat or green vegetables to balance their nutritional needs. Meat, vegetables, fruit, and fishery products are important sources of vitamins and micronutrients that are lacking in cereals. Developed nations have vitamin pills and enriched foods to make up for potential dietary shortages, but pills and enriched food are not yet practical options for the developing world. Specifically, lack of proper nutritional variety in African diets, in some cases along with inadequate calories, has resulted in 65 percent of African women of childbearing age being anemic, 40 million children being severely underweight, and 50 million being vitamin-A deficient.

Although food prices have become lower and many developing nations produce enough food in the aggregate, many people still go hungry. Indeed, food intake among the poorest people experienced negligible improvement throughout the Green Revolution (Lipton and Longhurst, 1989). This is particularly true in poor rural areas where food prices are still too high, and people cannot grow enough food or buy enough food because their income is too low or they are unable to find food in other ways. A large number of the poorest and most oppressed people are women. In Nepal, women work an average of 11 hours a day compared with seven hours a day for men. The schooling of women is also inadequate: For developing nations as a whole, there are 86 women for every 100 men in primary schooling and 75 for every 100 in secondary schooling (Conway, 1998).

The Green Revolution has fallen short of providing the sorts of nutritional changes that would relieve some of these specific problems, not to mention the problems surrounding food distribution to ensure that the increased food supply is getting to the places where people need it the most.

Socioeconomic Challenges

As a result of increased food production in the Green Revolution, a number of socioeconomic changes occurred in the regions where the revolution "succeeded." One result that occurred in the Indian Punjab was that the demand for land increased dramatically, driving up its price as much as five times. In many cases, only the wealthiest could afford to buy the land and often converted their tenants into hired laborers to reduce costs (Cleaver, 1972). This situation resulted in a greater dichotomy between the rich and the poor, and landowners and laborers. Thus, socioeconomic power became more concentrated in the hands of a few rather than many.

What are some of the causes of these inequities? In India, for example, adoption of Green Revolution technologies was strongly correlated with water supply. Where irrigation was available, Green Revolution adoption was nearly 100 percent, but where irrigation was unavailable, adoption was under 50 percent. For the most part, small farmers on less-well-favored lands received few benefits and in some cases became poorer, because the price of grain went down. In many cases, small farmers also suffered from loss of land ownership. In Ethiopia, for example, the Chilalo Agricultural Development Project that began in 1967 led to massive evictions of landowners' tenants as large farmers replaced human labor with tractors and combines. Before the project, tenancy rates were at about 50 percent; by 1972, tenancy had fallen to about 12 percent. Many of the displaced small farmers moved to urban areas (Conway, 1998).

These displaced farmers then moved into the cities and experienced little, if any, improvement in their living standards. Women in some settings lost access to and control of more primitive agricultural resources in the transformation to more modern agriculture. In Mexico, when the elimination of subsistence farming became government policy, women were not provided with new economic roles in the modern sector. Many moved into low-paying jobs in newly built factories, which provided neither the security nor the dignity of traditional agriculture (Perkins, 1997).

Not all socioeconomic changes led to greater disparities between the land-owning and the landless, however. The early adopters of HYVs were typically those with larger farms. Over time, however, small farmers adopted the HYVs as well, so that, in South India at least, there were no systematic differences in adoption rates by farm size (Hazell and Ramasamy, 1991). Moreover, work activities other than cultivation and agricultural wage work became more important in South India. Those activities included weaving, herding sheep, and a variety of service occupations, including local government jobs (Harriss, 1991).

The technologies introduced in Asia and Latin America in the Green Revolution generally required not more land, but chemical fertilizer and well-timed water. Farmers who could access these inputs did well while others did not. To the extent that large landholders also had access to fertilizer and irrigation, they tended to adopt the new technologies early and successfully.

It is important to remember that the goal of the Green Revolution was not to provide a cure for socioeconomic disparities, but to improve food production. Hence, these ancillary problems still exist and will persist until the emergence of a more comprehensive agricultural revolution makes socioeconomic well-being across all stakeholders a priority.

Environmental Challenges

While the Green Revolution played a key role in achieving food security and reducing rural poverty in key areas of the developing world, it was environmentally harmful in many settings. Some of the best-irrigated lands have become overly saline[3] since irrigation was introduced as part of Green Revolution technology, which has made crop planting difficult if not impossible. Irrigation cannot continue today in many areas because of decreasing water availability and resulting soil salinity. These conditions were exacerbated during the Green Revolution by the need for irrigated lands so that high-yield varieties could succeed (Ruttan, 1998; Conway, 1998). In addition, water tables have diminished, leading to problems with water accessibility for agriculture.

Moreover, waterways and soils have become contaminated by the large amounts of pesticides and fertilizers used on Green Revolution farmlands. Pesticides and nitrates in drinking water have proven to be detrimental to human and animal health. An unfortunate consequence of overuse of pesticides in particular areas is that crop pests have developed resistance to the pesticidal chemicals, rendering the chemicals ineffective. Indeed, these environmental spillovers have had a depressing effect on agricultural production, the very thing they were intended to improve (Ruttan, 1998).

Increasingly, agricultural planners are realizing the need for a "Doubly Green Revolution" (Conway, 1998)—one that will provide

---

[3] The water used in irrigation is usually saline seawater.

the kinds of yield increases achieved by the Green Revolution while being environmentally sustainable and protective of human and animal health.

Lessons from the Green Revolution

Although the world has changed since the mid-20th century, it is still possible to draw lessons from the Green Revolution's successes and failures to inform the GM crop movement. We believe that the following lessons, condensed from the analysis presented in this chapter, will enable future agricultural scientists and policymakers to learn from and build upon the Green Revolution's many successes, and at the same time to address the issues it left unsolved and to avoid its mistakes.

Successes of the Green Revolution

- In the developing world, a factor that is as important as yield improvement is consistency of yield, especially in the face of uncertain conditions such as sporadic rainfall and unanticipated pest infestations. The HYV seeds and accompanying chemicals of the Green Revolution were able to accomplish yield consistency. Genetically modified crops must also be able to provide such consistency of results.
- The Green Revolution demonstrated that public-sector funding and leadership are desirable, and may be necessary, to garner a wide base of financial support and make the costs of agricultural technology reasonable to growers. Likewise, it may be beneficial for the GM crop movement to gain public support for a successful transfer to the developing world.
- Vital to the success of the Green Revolution was the practice of training scientists and farmers in the developing world so that they would be able to carry out the farming techniques independently and to train others in their localities—in other words, the promotion of sustainable agricultural practices. Likewise, if

GM crops are to be a benefit to the developing world, local scientists and farmers must learn how to use them appropriately.

- Farmers must be able to afford the technologies. Green Revolution methods to achieving this end, such as the loan systems described earlier in the "Funding" section, proved to be successful and may be worthwhile for GM crop adoption in the developing world.

- Local infrastructures need to be conducive to the introduction of key technologies. The fact that this was the case in Latin America and Asia, but not in Africa, and that it made a difference in where the Green Revolution really took off point to the importance of appropriate infrastructures to support agricultural technology. Infrastructural considerations are equally important for the GM crop movement.

- During the Green Revolution, national governments had motivations other than the desire to increase food production behind their support for agricultural technology. For example, some countries, such as Mexico, were looking to completely reshape their national economy. Nations that are similarly motivated may be more open to accepting agricultural movements, such as the GM crop movement, which can secondarily bolster their other goals.

Issues Left Unresolved by the Green Revolution

- The Green Revolution's success in the developing world was measured by farmers' adoption levels and evidence of increased food production, rather than by considerations such as improved environmental quality or socioeconomic improvements. However, a new agricultural revolution will fall under public scrutiny as to whether adequate food distribution is achieved and what sorts of environmental or socioeconomic disruptions may take place. As such, environmental quality and socioeconomic improvements need to be taken into consideration. Meeting these concerns must be a priority for the GM crop movement.

- A new agricultural revolution, such as the GM crop movement, will need to address sub-Saharan Africa's unique agricultural challenges, because this is the area of the world that is most in need of improved food production and food-supply stability.

All the factors discussed in this chapter—new science and technologies, funding sources, regions of interest having appropriate local infrastructures (so that the movement can take root), political motivations, and economic policies—must fall into place in the right way for an agricultural revolution to succeed. If GM crop technology is to achieve revolutionary status on a global scale, these factors would need to be aligned somewhat differently than they are at present. In the next chapter, we examine how these conditions must be aligned.

CHAPTER THREE

The Gene Revolution: Genetically Modified Crops

Since the mid-1980s, research teams in biotechnology firms world-wide have been transplanting genes across species to produce engineered crops with pest resistance, herbicide tolerance, tolerance to drought and saline soils, and enhanced micronutrient content. These genetically modified crops were first commercialized on a wide scale in the early 1990s. Today, they make up anywhere from a quarter to three-quarters of the total acreage of select crops in the United States, Canada, Argentina, and China (James, 2003).

This new technology comes at a time of great need for increased food production in certain regions of the world. In its heyday, the Green Revolution achieved dramatic successes, transforming agricultural production and averting wide-scale famine in many parts of the world. But the transformation it engendered was not complete. As we discussed in Chapter Two, despite all that the Green Revolution accomplished, hunger and malnutrition have persisted, particularly in Africa. Food security, the long-run sustainability of agricultural production systems, and the quality of the natural resource base are still important issues worldwide. In addition, pressures from population growth and inappropriate socioeconomic infrastructures have created problems with deforestation, soil erosion, and pollution (Alston, Norton, and Pardey, 1995).

All of these issues call out for a new agricultural revolution. The Green Revolution laid out a path for what needs to happen for a new agricultural movement to be revolutionary. In those terms, the GM crop movement, or "Gene Revolution," may have great potential to be revolutionary because it has already reached some of the points on that path. Indeed, it can already be called a revolution on a limited scale. But there are significant differences between the Green and Gene Revolutions—both in the technology they employ and the context in which they exist.

In this chapter, we describe what is currently happening in the GM crop movement, what is *not* happening that should happen if it is to become revolutionary in the developing world, and the significant obstacles that stand in the way of its success. We again closely examine certain aspects of the movement discussed in Chapter Two: science and technology, funding, areas where the movement is occurring, and policies and politics.

Science and Technology

Like the Green Revolution that preceded it, the GM crop movement employs a combination of previously unheard-of technologies in plant breeding, with the goal of improving agricultural yields. Unlike the Green Revolution techniques, Gene Revolution techniques involve cutting-edge biotechnological advances to achieve this goal. And for the most part, scientists in the developing world have not yet been trained in these technologies such that they can produce their own GM crop seeds.

The term "biotechnology" encompasses all the techniques that use organisms, or parts of organisms, to produce or alter a product, or that develop microorganisms for specific purposes. By far the most prominent example of biotechnology in agriculture is genetically modified crops. GM crops contain genes that are artificially inserted instead of the plants' acquiring them through sexual means. The *transgene* (artificially inserted gene) may come from a completely different species. Bt corn and Bt cotton, for example, contain genes from different subspecies of a soil bacterium *Bacillus thuringiensis* (Bt) that give plants the ability to manufacture their own pesticides. Depending on where and for what purpose the plant is grown, desirable genes may provide features such as higher yield or improved quality;

pest or disease resistance (as with Bt crops); tolerance to heat, cold, or drought; or enhanced nutritional content—all desirable agricultural traits for farmers and consumers in the developing world.

Agricultural Benefits of Genetically Modified Crops

Many genetically modified food crops are now planted around the world. The most common are GM soybeans, followed by GM corn, cotton, and canola. Others are GM tomatoes, potatoes, papayas, chicory, melons, rice, squash, sugar beets, and wheat.[1]

One of the most common and important purposes of GM crops today is to confer tolerance to herbicides that are sprayed on farmland to control weeds. Herbicide-tolerant crops include transgenes providing tolerance to the herbicides glyphosate or glufosinate ammonium. These herbicides are broad spectrum, meaning that they kill nearly all kinds of plants *except* those that have the tolerance gene. Thus, a farmer can apply a single herbicide to his fields of herbicide-tolerant crops, and can use the herbicide effectively at most crop growth stages as needed. Another important benefit is that this class of herbicides breaks down quickly in the soil, eliminating residue carryover problems and reducing adverse environmental impacts (Byrne et al., 2004).

Another common purpose of genetic modification of crops is to confer protection against insect pests. Importantly, this crop trait could substantially improve yields in the developing world where pest damage is rampant and/or reduce use of chemical pesticides. The soil bacterium *Bacillus thuringiensis* in Bt corn and Bt cotton produces crystal proteins that are toxic to certain insects but generally harmless to vertebrates and non-lepidopteran insects. The genetic insertion of the bacterial gene into the plant genome enables the plant to produce its own pesticide. Depending on the subspecies of the bacterium from which the gene is taken, the pesticide is toxic to insects of the orders *Lepidoptera* (which includes the common corn pests, European corn borer, Southwestern corn borer, and corn earworm), *Diptera* (mos-

---

[1] See www.agbios.com for more information.

quitoes), or *Coleoptera* (beetles) (Wu, 2002). Other purposes of modification in current and past GM crops include changing the fat and acid composition of the crop (canola and soybeans), conferring male sterility (chicory), conferring resistance to viral infection (papaya), and delaying time to ripeness to allow for longer transportation time and shelf life (tomatoes).

Future GM crops may be able to benefit farmers in other ways previously unrealized. Some GM crops may be able to grow in conditions that have been unsuited for agriculture. For example, scientists have developed and may soon be marketing a type of genetically modified tomato that is able to grow in salty soil. This type of modified tomato is particularly useful in areas where crops could not be grown previously because of soil salinity. The tomato itself channels the excess salt into its leaves so that the fruit retains its "normal" flavor (Zhang and Blumwald, 2001). Other GM crops are now being developed that are well suited to drought conditions, that survive in soil heavy in metals such as aluminum, that convert or "fix" nitrogen from the air, and that produce vaccines against common diseases such as cholera and hepatitis B (Byrne et al., 2004). Because farmers would thus be able to grow crops in otherwise difficult conditions, these crops, if affordable, could be an attractive option for farmers in areas with poor agricultural lands, notably in sub-Saharan Africa.

Potential Health Benefits of GM Crops

Most of the types of genetically modified crops grown today were produced with the intent to benefit agriculture, in the variety of ways described above. However, even these current types of GM crops have some indirect health benefits. For example, crops such as Bt corn that are genetically modified to produce insecticidal proteins have been shown in field studies to have lower levels of *mycotoxins*— chemical toxins and carcinogens produced by fungi (Munkvold and Hellmich, 1999; Dowd, 2001; Schaafsma et al., 2002; Hammond et al., 2003). While the benefits of mycotoxin reduction in Bt crops are achieved primarily through reduced market losses in the industrialized world (i.e., farmers' crops will be more readily accepted if mycotoxin contamination is low), this benefit could have a significant health impact in the developing world, where exposure to food-borne mycotoxins is higher and few regulations exist to protect consumers (Wu, Miller, and Casman, forthcoming). Moreover, because GM crops are more cost-effective for the farmers that adopt them, food prices go down, resulting in savings to consumers (Conway, 1998; Wu, 2004). Decreased food prices could then lead people to buy a greater variety of foods (Conway, 1998), thereby improving the nutritional content of their diets.

Future GM crops could also have substantial direct nutritional or medicinal benefits to consumers. Crops could, for example, be genetically modified to produce micronutrients vital to the human diet. One type of crop in the making is "golden rice," genetically modified to produce beta-carotene, the precursor to vitamin A. Currently, about 400 million people in the world are at risk of vitamin A deficiency, which can lead to serious morbidity including night blindness, respiratory diseases, and even childhood death (Toenniessen, 2000). Rice modified with daffodil and bacterial genes would be able to produce beta carotene, which could in turn lead to a more balanced diet that could contribute to enhanced health in regions where rice is a staple of the diet. This type of crop would thus be potentially beneficial among Asian and African populations that suffer from malnutrition. Canola, too, can be genetically modified to enhance vitamin E content or to better balance fatty acids (Byrne et al., 2004).

Food crops engineered to produce edible vaccines against infectious diseases would make vaccination more readily available to children around the world. Because of their palatability and adaptation to tropical and subtropical environments, bananas have received considerable research attention as a vehicle for vaccine delivery. Transgenic (i.e., GM) bananas containing inactivated viruses protecting against common developing-world diseases, such as cholera, hepatitis B, and diarrhea, have been produced and are currently undergoing evaluation (Byrne et al., 2004). Because they would produce only the necessary antigens, these types of vaccine-producing GM crops may be safer than traditional vaccines whose additional materials often cause harmful side effects (Conway, 1998).

Table 3.1 summarizes current and potential environmental and health benefits of genetically modified crops. The future may bring even more benefits as genetic technologies improve.

Potential Risks of GM Crops
In the first decades of the Green Revolution, risks to human health and to the environment from the accompanying pesticides, fertilizers, and irrigation were not a significant concern. Consequently, the Green Revolution had little problem achieving the level of public acceptance that was necessary for it to have a revolutionary impact. But GM crops are a different story, due in part to the adverse environmental outcomes of the Green Revolution that eventually came under public scrutiny.

The introduction of genetically modified crops into the food supply has generated a number of concerns about potential risks associated with this new agricultural technology. Depending on the seriousness of these risks, they may limit the GM movement's impact. The risks fall under the broad categories of human health and the environment. When a new gene producing a novel protein is introduced into a crop, there is a chance that human subpopulations may have an allergic reaction to the protein (Byrne et al., 2004). Also, GM crops may adversely affect non-target species; for example, a GM pest-protected plant targeting lepidopteran pests may spread toxin-

Table 3.1
Current and Potential Benefits of Genetically Modified Crops

Agricultural Benefits	Human Health Benefits
Herbicide tolerance	Reduction of mycotoxin contamination
Protection against insect damage	Lowered food costs potentially leading to more-varied diets
Virus resistance	
Tolerance to salty soil	Nutrient enhancement
Drought tolerance	Vaccine production

containing tissues, such as pollen, that may contaminate the food of non-target lepidopterans.[2] Cross-pollination between GM crops and non-GM crops, or GM crops and wild plant species, may occur, with unknown consequences. Insects may eventually develop resistance to the insecticidal proteins produced by Bt crops, leading to inefficacy of both these crops and microbial Bt sprays (the latter of which is valued among organic farmers) (Gould et al., 1997; Andow and Alstad, 1998). A host of other risks surrounding GM crops may exist of which we are currently unaware (Wu, 2002).

Many of these risks, particularly those regarding food safety and impacts on non-target species, have undergone extensive scientific research (U.S. Environmental Protection Agency [EPA], 2001). Most of this research has found no evidence that such risks exist among current GM crops. Indeed, 81 scientific studies financed by the European Commission have all shown no evidence of risk to human and animal health or to the environment from genetically modified crops (Paarlberg, 2003). However, it is impossible at this stage to fully investigate all potential risks, current and future, of GM crops, and the challenges posed for such investigations by the ongoing development of GM varieties are substantial.

Funding

The Green Revolution has made it clear how important funding is to enable an agricultural movement to become revolutionary. In that case, financial support came largely from the public sector. Today, there is still a significant need to fund new agricultural research and technology transfer in the developing world. But biotech research and development (R&D) takes place almost exclusively in the developed world within private industry.

Genetically modified crops are largely the product of private industry. This is partly because new technologies are far more costly

---

[2] This particular risk has been shown to be insignificant in the United States (Sears et al., 2001), but has not been thoroughly tested in the developing world.

than existing ones, and the biotechnology industry was able to gather the necessary funds to develop these technologies long before public awareness of GM crops could lead to publicly generated funding for GM crop development (Pinstrup-Andersen and Schioler, 2001). Successful companies must focus on their markets with the intent of generating profit. With regard to agricultural biotechnology, companies in the United States and elsewhere have thus far created primarily seeds that farmers in industrialized countries can and will purchase: corn and soybeans that can tolerate a particular herbicide, corn and cotton that are resistant to particular pests, and food crops that last longer on the supermarket shelf. Because of the "technology fee" that growers pay to use these crop seeds (including recoupment of industry's research and development costs as well as profit), and because the seeds are designed particularly for their planting situations, the targeted farmers in industrialized countries have generally found it worthwhile to buy these seeds and have been willing to pay the technology fee (Wu, 2004).

For this reason, there are currently two primary financial obstacles to using agricultural biotechnology in the developing world: (1) the lack of incentive on the part of the biotechnology industry to produce GM crops that are applicable in the developing world; and (2) intellectual property issues that discourage R&D by organizations, public and private, other than the firm first developing a technology. Because of these obstacles, GM crops that would be important to farmers in the developing world are in many places either unavailable or too expensive to be of practical use. Unless these crop seeds can be made more affordable, farmers will lack the incentive to use them and the chance of a revolutionary impact in the developing world will be small.

A fundamental challenge in this newest agricultural movement that did not arise during the Green Revolution is the definition and treatment of intellectual property (IP). IP issues are central to the Gene Revolution because whereas science and technology move forward through the sharing of ideas and resources, IP ambiguities and restrictions can often limit the valuable diffusion of science and technology. Commercial application of biotechnology has taken place primarily in the United States and primarily through the private sector. The issue of who "owns" a particular *event* (the successful transformation) of a genetically modified crop and who can develop it further has become so economically important and contentious that numerous cases involving this issue are now being litigated (Woodward, 2003). Some observers consider IP issues to be one of the most important impediments to the development and adoption of GM crops in the developing world (Shoemaker et al., 2001; Cayford, 2004).

Several large multinational corporations that combine seed industries with chemical industries currently dominate agricultural biotechnology. These profit-driven companies have seen little reason to invest in expensive research and regulatory costs to produce crops that are grown on relatively few acres and must be heavily subsidized for poor farmers to afford. Inevitably, private research focuses on needs of capital-intensive farming; research to feed the poor is less attractive, as it involves long lead times, risk related to unpredictable agricultural conditions and disregard of intellectual property rights, and beneficiaries with little capacity to pay (Conway, 1998). New agricultural technologies are usually expensive to develop because of the need for a large research infrastructure and the cost of meeting strict regulatory requirements. Therefore, corporations have focused their resources on creating biotech crops for farmers in industrialized nations, and not on biotech varieties of developing-world "orphan crops"[3] such as cassava, millet, sorghum, cowpeas, or yams (Conway, 2003).

Because of the private sector's role in the financial organization of the GM crop movement, IP considerations arise. Corporations own the rights to many of the necessary technologies and knowledge that lead to the development of useful GM crops; thus, public researchers in agricultural biotechnology as well as other private companies are often unable to access those technologies and knowledge that they need, or are legally blocked from using what they do know.

---

[3] Orphan crops, while not considered important as far as export markets (as are maize, wheat, and beans), are nevertheless important in the diets of people in certain less-developed countries.

Yet, it is important to remember that lack of intellectual property protection could make industry less willing to invest in developing countries (Shoemaker et al., 2001).

This is not to say that, left to its own devices, the private sector would ignore the concerns of the developing world altogether when developing agricultural biotechnologies. In fact, some firms have recently expressed an interest in working with regional institutions to develop crops that would be beneficial to the developing world. In addition, they are willing to donate a substantial portion of their scientific knowledge, such as genomes of key food crops, for the purpose of increasing agricultural knowledge capital in the developing world.

Moreover, several public-sector initiatives or partnerships between private and public sectors have emerged, with the technology transfer of GM crops to the developing world as one of their goals. A notable example is the Rockefeller Foundation's African Agricultural Technology Foundation (AATF), which seeks to facilitate useful combinations of biotechnology with other methods of improving and empowering African agriculture. AATF is an African-led, African-based organization designed to eliminate many of the barriers that have prevented smallholding farmers from gaining access to agricultural technologies. Genetically modified crops, as well as other tools of biotechnology, are one part of this goal; but importantly, AATF works toward an integrated approach to solving food security issues in Africa, which was previously left largely untouched by the benefits of the Green Revolution. Another such example is the International Service for the Acquisition of Agri-biotech Applications (ISAAA). Supported by a combination of corporations, government agencies, and foundations, ISAAA is a nonprofit organization that focuses on bringing GM crops to the developing world.

However, studies on industrial decisionmaking in agriculture have shown that key determinants of the private sector's decision to invest in agricultural research are: (1) the perceived size of the market, and (2) the ability to reduce transaction costs when farms are larger (Shoemaker et al., 2001). Most markets and farms in the developing world are small, and thus provide little incentive to the private sector to fund agricultural research that will help them.

Several important funding practices would need to occur for the GM crop movement to become revolutionary in the developing world. As the history of the Green Revolution makes clear, publicly funded research, if deliberately aimed at low-cost food production, can benefit developing-world farmers (Conway, 1998). Thus, one important component of the GM crop movement in the developing world is increased public-private-sector partnership. While this is happening to a small extent in the GM crop movement today, increased collaboration between these sectors would likely expedite the transfer of agricultural biotechnologies to poorer regions of the world. Another important component of the GM crop movement is that intellectual property processes must be established to ensure appropriate protection and incentives to laboratories and manufacturers, while providing public researchers and developing-world farmers with access to the necessary tools for technology transfer. Such processes will allow GM crop seed to be affordable to the growers that would most benefit from them.

Where the Gene Revolution Is Occurring

The Green Revolution had such a sizable impact because it took place in areas that were ripe for an agricultural revolution—namely, Asia and Latin America in the developing world and Great Britain among industrialized nations. Concerns about impending massive malnutrition, food security in the near future, and unwelcome political influences were all key pressures driving the need to adopt Green Revolution technology. Today, such needs are still present in much of the developing world, and most of all in sub-Saharan Africa. How has the GM crop movement been addressing these concerns?

Thus far, the Gene Revolution has made its mark in only several places. Four nations—the United States, Canada, Argentina, and China—are together planting 99 percent of the world's total acreage of genetically modified crops (James, 2003). China, in fact, was the first country to commercialize GM crops in the early 1990s, with the introduction of virus-resistant tobacco, followed soon by virus-resistant tomatoes (James, 1997). Some five million farmers have been growing GM cotton there for six years, with higher yields and without laborious and hazardous pesticide spraying (Conway, 2003). And as of March 2004, Britain has allowed the commercial planting of one type of GM herbicide-resistant corn.

It is too early to forecast where around the world genetically modified crops will provide a benefit to local agriculture and where they might be adopted. After all, GM crops have been commercialized for only about a decade, while Green Revolution technologies have been in use more than four decades. Even though there is a clear need for radical improvement in food production in certain regions, real obstacles stand in the way for agricultural biotechnology to meet that need. One obvious constraint is that the types of GM crops available today are not beneficial in many parts of the developing world. For example, the genetically modified herbicide-tolerant crops that are so popular in the United States and Canada are of little use in developing nations where farmers do not use herbicides on a regular basis. Another constraint, discussed earlier, is that the new agricultural biotechnologies that may be beneficial are currently unaffordable for many farmers in the developing world.

Figure 3.1 shows that while the total acreage devoted to GM crops is steadily increasing in both industrialized and developing countries, GM crop acreage is much higher in industrialized areas. This is despite the considerably larger acreage of farmland in the developing world.

Of those parts of the developing world where GM crops are grown, South Africa and India are the most prominent. In the Makhathini Flats of South Africa, for example, farmers have been growing GM cotton for four years (Conway, 2003). India, a nation that had previously shunned agricultural biotechnology, recently harvested its first Bt cotton crop (genetically modified to protect against insect pest damage). Qaim and Zilberman (2003) found a 79 percent average increase in Bt cotton varieties over conventionally planted varieties in India in 2002. If GM crop adoption in these nations is sustained over the next several years and possibly beyond, the GM movement may indeed be considered revolutionary in those areas.

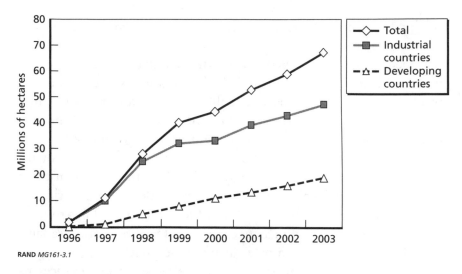

RAND *MG161-3.1*

SOURCE: James, 2003.

Figure 3.1—Area Devoted to Genetically Modified Crops, 1996 to 2003

Closely linked to the question of whether the GM crop movement becomes revolutionary in a particular region is whether the people of that region are willing to accept the new technologies and to adapt their lifestyles to meet the changes that the technologies may bring—criteria for a successful agricultural revolution.

In terms of a people's acceptance of the new technologies, first, there are the direct impacts on food and culture. Compared with most Americans, people in many parts of the developing world place a higher level of importance on food as part of their culture (Echols, 2002). Americans themselves felt wary about hybrid corn when it was first grown on U.S. farms in the 1920s and 1930s (Griliches, 1957), so it is not difficult to imagine what the cultural impact might have been, or might be, in nations worldwide from the high-yield hybrid crop varieties of the Green Revolution, and even more so from genetically modified crops. To this end, GM crops have already become a stigmatized technology in many parts of the world (Paarlberg, 2002) because of a widespread fear of "Franken-food" and "playing

God" with nature. Mistrust of industry may also play a role in these concerns.

Second, GM crops may have indirect impacts that change the culture of a region more slowly, which will affect whether the GM crop movement can become sustainable in a revolutionary way. Shifts in socioeconomic status due to increased food production and new agricultural methods can benefit certain groups in society but may cause other groups to suffer. For example, in the first third of the 20th century, machines increasingly replaced humans on farms, first in the industrialized world, then in parts of the developing world. But only those farmers who had access to sufficient land holdings to make the machinery useful had incentives to participate, and of those farmers, only those who could afford the purchase price could participate. Smaller farmers tended to suffer from the lower product prices resulting from increased production overall, and many sold or lost their land. GM crops, besides being costly, may also favor those farmers with large land holdings—an outcome that would be unacceptable to many stakeholder groups. The lack of acceptability of this outcome on a regional basis, combined with the price of GM crops, could prevent the GM crop movement from achieving revolutionary impacts in some regions.

Moreover, the new intellectual property issues associated with GM crops would require agricultural practices that differ substantially from traditional farming methods among subsistence farmers. Patent rights require farmers who want to plant GM crops to buy new seed every season. This requirement is not usually a problem in the industrialized world, because industrialized-world farmers have already been buying new seed for each season since their adoption of hybrid varieties. Agribusiness in richer nations values crops that have uniform qualities (e.g., uniform height, weight, and development time), and these qualities are much more easily achieved through seed newly bought each year (Pinstrup-Andersen and Schioler, 2001). But developing-world farmers are often not willing or able to buy new seed every season. About 80 percent of developing-world farmers currently save seed from the previous season (Pinstrup-Andersen and Schioler, 2001) in order to cut costs. Under current intellectual property standards, a substantial change not just in practice, but also in the shift of cultural mindset from subsistence living to agribusiness, would need to take place for the Gene Revolution to succeed among developing world farmers.

Policies and Politics

Interestingly, the main constraint on the potential of GM crops in the developing world may not have to do with either agricultural conditions or credit availability, but with precautionary regulations. Policies and regulations for genetically modified crops have evolved very differently from those for Green Revolution technologies. In the years leading up to the Green Revolution, policymakers saw a political need for agricultural improvement worldwide, advanced this cause as a key policy concern, and ensured that funds were appropriated toward the cause. Plant scientists moved into foreign nations and aided regional farmers in adopting new agricultural technologies almost completely uninhibited by governmental regulations. Compare this with the current Gene Revolution: Biotech industries created genetically modified crops, and governmental agencies, encouraged by scientists at a 1973 Gordon Research Conference (Wu, 2002),[4] saw the need to regulate these GM crops to prevent or mitigate potential health and environmental risks. Now regulations, both domestic and international, and both directly and indirectly, are key in determining whether or not this Gene Revolution will spread to particular regions of the world.

As of yet, there does not appear to be a strong political motivation for genetically modified crops to succeed in the developing world. Communism is no longer a threat, and famines, while still a problem in parts of the world, appear to be more the result of localized weather, politics, and war conditions than a sweeping threat that

---

[4] Gordon Research Conferences are among the most prestigious scientific conferences in the world, bringing together top international experts on a variety of topics in biology, chemistry, and physics.

commands sustained government and public attention in industrial-ized countries. Instead, public concerns and national and interna-tional regulations are now the driving force behind whether GM crops are adopted or rejected in various parts of the world, because wider public scrutiny and the newness of the science have led to con-cerns about environmental and health risks of GM crops that must be dealt with at the policy level. The battle between U.S. and European Union regulations, which feature very different stances on the accep-tance of GM crops in food and feed, has been a major determinant of this outcome.

The regulation of GM crops worldwide has split into two major camps, with some nations strongly advocating for GM crops (led by the United States) and others strongly advocating against them (led by the European Union [EU]). The regulations in these two parts of the world, and battle between these two factions over the place of GM crops in global food production, are shaping the regulations in other nations worldwide. Hence, the discussion in this section focuses on the current institutional framework by which policy decisions on GM crops are made in the United States and the EU.[5]

United States
For the purposes of regulating the technology of genetically modified organisms (GMOs), a Coordinated Framework for Regulation of Biotechnology (51 *Federal Register* at 23302, 1986) was created within the Office of Science and Technology Policy (OSTP) in 1986. However, its outcome was not so much to form new regulations for GMOs as to delegate responsibility for oversight to existing agencies under the framework of existing statutes. Importantly, the Coordi-nated Framework emphasized that risk assessment would be based on the biotech *product* itself, not the *process* by which the product was developed. This regulatory approach, which has proven to be more permissive toward GM technologies with the direct impact of aiding

---

[5] Useful oversights of regulations in other individual nations are given in Nap et al. (2003) and Paarlberg (2001).

the Gene Revolution in the United States, is quite different from the European Union approach, which emphasizes the process.

Under this framework, GM crops are subject to the statutes of the Federal Food, Drug, and Cosmetic Act (FFDCA); the Federal Plant Pest Act (FPPA), now modified into the Plant Protection Act (PPA); the National Environmental Policy Act (NEPA); the Federal Plant Quarantine Act (FPQA); and the Federal Insecticide, Fungicide, and Rodenticide Act (FIFRA). Hence, GM crops are regulated by three governmental agencies: the United States Department of Agriculture, the U.S. Food and Drug Administration (FDA), and the Environmental Protection Agency (EPA). Table 3.2 depicts the regulatory scheme for coordinating reviews of GM crops.

U.S. Department of Agriculture. Usually, the first agency to evaluate the safety of new transgenic crops is the USDA. The Animal and Plant Health Inspection Service (APHIS) of the USDA has regulatory oversight for protecting U.S. agriculture from pests and diseases. There are two main steps to the GM crop approval process, one for field tests and one for commercialization, under APHIS regulation. First, registrants must obtain a permit to test the crop on controlled fields. APHIS issues permits after it has concluded that the GM plant in question exhibits no plant pathogenic properties; is no more likely to become a weed than its non-engineered parental varieties; is unlikely to increase the weediness potential for any other cultivated plant or native wild species with which the organism can inter-

Table 3.2
Regulatory Scheme for Coordinating Reviews of GM Crops

Product Class	Lead Agency	Federal Statutes
Plants	USDA/APHIS	FPPA, NEPA, PPA, FPQA
Pesticides	EPA	FIFRA, FFDCA, FPQA
Food and additives	FDA	FFDCA

SOURCE: National Academy of Sciences, 2000a.

breed; does not cause physical damage to processed agricultural commodities; and is unlikely to harm other organisms, such as bees, that are beneficial to agriculture (Carpenter, 2001). Second, to commercialize a transgenic plant, the researcher petitions APHIS for non-regulated status—a status that has aided the Gene Revolution in the United States by its relief from regulatory restrictions from the USDA, except in the unusual case of proven harm.

U.S. Food and Drug Administration. While the FDA has broad regulatory authority to determine the safety of foods, food ingredients, and feed through the FFDCA, there is no particular statutory requirement regarding regulation of *genetically modified* foods. This is largely the result of the FDA's 1992 policy statement (57 *Federal Register* at 22991, 1992). In this policy, the FDA granted Generally Recognized As Safe (GRAS) status to GM foods, which meant that these foods did not need to undergo additional tests or analysis with the agency unless there was a claim of a nutritional or health benefit. Thus, the FDA recommends, but does not require, that biotech industries producing GM crops undergo consultation with agency scientists before marketing their products (U.S. Congress 1958, Section 402(a)(1)).[6] This ruling has greatly decreased the regulatory burden that GM crops undergo in the United States, again contributing to the ease with which the Gene Revolution has spread.

Another action that has aided the Gene Revolution in the United States, but has caused controversy elsewhere in the world, was the FDA decision that GM foods do not require labeling (57 *Federal Register* at 22991, 1992). The FDA currently requires labels on GM products only if the genetic modification introduces a protein from a food that is a known allergen (e.g., peanuts). However, because proteins produced by GMOs currently on the market are not known allergens, they need not be labeled under FDA policy. Furthermore, the FDA states that there is no basis to distinguish GM foods as a class separate from foods developed through traditional breeding.

---

[6] In January 2001, however, the FDA proposed to make pre-market consultation mandatory (66 *Federal Register* at 4706, 2001).

Environmental Protection Agency. By the authority of the Federal Insecticide, Fungicide and Rodenticide Act (7 U.S.C. Section 136, 1972), the EPA regulates the sale, distribution, and use of pesticides, including those produced by GMOs. To be registered under FIFRA, a pesticide must not cause "unreasonable adverse effects on the environment." This statement has also facilitated the greater spread of the Gene Revolution in the United States as compared with the European Union because of the relatively lenient burden of proof. That is, GM crops are registered if preliminary tests show no evidence of unreasonable adverse effects.

As with the USDA's process, there are several steps involved in the EPA's regulatory process for GM crops. The first step for registrants or researchers is to obtain EPA approval for an "experimental use permit" for field trials. It is also possible for the registrants to apply for an exemption from the requirement for a demonstrated tolerance for their products (under FFDCA). To qualify for an exemption, the product must be shown to be nontoxic to mammals (including humans) and to have little potential for allergenicity in humans, based on in vitro digestibility tests, thermolability, and amino acid sequence comparison with an allergen database. In the final stage prior to full commercialization of a product, the EPA reviews the application for registration (FIFRA, Section 3) on the basis of the following areas of required data:[7] product characterization, human health assessment, ecological assessment, insect resistance management, and benefits (EPA, 2001).

European Union

The European Union has adopted a very different regulatory approach from that of the United States to deal with GM food and feed products coming to market. In 1990, the European Economic Community (EEC) issued "Directive 90/220 on the Deliberate Release to the Environment of Genetically Modified Organisms," under the agency Directorate General (DG) XI for Environmental Protec-

---

[7] In most instances, some or all of these data are known or in place prior to an experimental use permit.

tion (European Economic Community, 1990). Three key reasons for the regulation were given in the preamble to Directive 90/220:

1. That precautionary action should be taken with regard to the environment,
2. That protection of human health and the environment requires that any risks from the release of GMOs to the environment should be controlled, and
3. That any variation between the member states concerning these rules would have an adverse effect on the operation of a single market.

In essence, the Precautionary Principle, which states in its most basic form that when an activity raises threats of harm to human health or the environment precautionary measures should be taken even if some cause-and-effect relationships are not established scientifically, became an explicit element of European regulation of GMOs, in contrast to the U.S. regulatory philosophy, which centers on evidence of unreasonable adverse effects. This Precautionary Principle has thus far halted the spread of the Gene Revolution in Europe, because of the inherent difficulty of proving that there is no risk involved with GM crops.

Moreover, Directive 90/220 enables member states to adopt stronger measures if they wish to do so. Some member states have created their own safety plans for GMOs, going far beyond the EU statutory basis. For example, even after the EU approved Bt corn for registration, Luxembourg and Austria banned it, stating that Bt corn was an environmental and commercial threat to organic agriculture. When the European Commission (EC)[8] attempted to remove these bans, only a few member states supported this attempt; therefore, the bans stayed in place (Levidow, 1999). To this day, the GM crop

---

[8] The European Commission acts as the EU's executive body. The 25 members (as of this writing) of the EC are drawn from the various EU countries, but they each swear an oath of independence.

movement has been blocked or delayed by individual countries despite EU approval; hence, there exists a de facto moratorium throughout Europe on the commercial growth, sale, and distribution of GM crops.

Part of the restrictions on GM crops in Europe stems from lawmakers' perception of strong public opinion against foods made from genetically modified materials and their perception that the public has a general lack of faith in the integrity of government regulators. This perception has been a limiting factor on biotechnological research and development. These anti-GM sentiments may be partly the result of other unrelated food crises that occurred in Europe in the mid-to-late 1990s, such as the outbreaks of "mad cow disease" (bovine spongiform encephalopathy) and more recently hoof-and-mouth disease (Laget and Cantley, 2001; Löfstedt and Vogel, 2001).[9] Unfortunately for U.S. agriculture, the purported anti-GM sentiments in Europe have also led to a marked decrease in agricultural exports to the EU since 1996, when Bt corn was first commercially planted in the United States (Wu, 2002). These sentiments also indicate that it is unlikely that GM crops will prove to be revolutionary to agriculture in Europe, due to both the lack of acceptability among its citizens as well as lack of support from major governing bodies.

The U.S. and EU Dispute over GMOs and Its Implications for the Gene Revolution in the Developing World

Recently, conflict between the United States and the EU regarding trade of GM crops has taken a new turn. With EU Directive 2001/18/EC,[10] which came into effect on October 17, 2002, and a subsequent European Parliament ruling on July 7, 2003, the European Union ended its unofficial moratorium on GM crops by passing

---

[9] Recent polls, however, revealed that EU public opinion is more nuanced and less unanimous in its viewpoint on genetically modified foods than what lawmakers' perceptions would suggest (Marris et al., 2002).

[10] Directive 2001/18/EC of the European Parliament and of the Council of 12 March 2001 on the Deliberate Release into the Environment of Genetically Modified Organisms and Repealing Council Directive 90/220/EEC (*Official Journal*, 2001, pp. 0001–0039).

new legislation regarding the deliberate release of GMOs into the environment. Those rules came into force on April 18, 2004. Under the new rules, everything from breakfast cereals to animal feed with more than 0.9 percent genetically modified content will need to be properly labeled, which requires carefully tracing GM foods through the entire food system.

Directive 2001/18/EC opens the way for biotech companies to apply for approval to market their products in EU countries, but American officials and others are doubtful that these rules will truly facilitate market access or encourage farmers to cultivate GM crops. Tracing and labeling GM food requires grain segregation at the farmer and elevator level. Not only would farmers and grain elevator operators need to keep GM and non-GM crops separated, they would also have to prevent commingling of the two types of crops during harvest, transport, and storage, which would likely slow the rate of turnover in a high-volume business.

Currently, elevator operators have very thin profit margins, and their profits depend on moving large volumes of product quickly (Lin, Chambers, and Harwood, 2000). To use the example of genetically modified Bt corn, if *all* U.S. grain elevators were to adopt segregation of Bt/non-Bt corn, the total segregation costs across the United States would be expected to exceed $400 million annually. This amount is higher than the total estimated monetary benefits of producing Bt corn in the United States, even taking into account potential environmental benefits from decreased pesticide use and health benefits from mycotoxin reduction (Wu, 2004). If these crop-segregation practices were to become the worldwide norm, even industrialized-world farmers soon may lose the incentive to plant GM crops because of high segregation costs, thus halting the spread of the Gene Revolution even in nations where farmers can afford the GM seed.

While these costs are unappealing to American farmers and food processors, they are an even more serious problem in developing nations that lack the infrastructure to segregate GM from non-GM crops in the first place. This problem came to a head in 2002, when worldwide wariness of GM crops received a level of attention that the

United States could not ignore. For instance, although Zambia was facing a serious famine, Zambian officials decided to turn away 26,000 tons of U.S. food aid in October 2002, saying that the shipments contained genetically modified corn that was unsafe. A key reason for refusing the aid, according to Zambia's agriculture minister, was that the GM corn seed could pollute the country's seed stock and hurt its export markets. In fact, government officials in Zambia, Zimbabwe, Mozambique, and Malawi all feared that if some of the unmilled GM corn imported as food aid was instead planted by farmers their nations would lose their current status as "GM-free" countries, compromising their ability in the future to export any food and farm products to the European Union, or to receive aid from European donors (King, 2002; Paarlberg, 2003; Wu, 2004). Hence, the Gene Revolution has ground to a halt in these nations due to potential fears of anti-GM policies abroad.

Zambia is not alone in its reluctance to allow any form of GM food—processed or unprocessed—through its borders. Surprisingly, China and India have also taken an anti-GM stance, despite past developments that had indicated their acceptance of GM crops. In 2002, India approved the planting of Bt cotton for the first time, and many expected that this movement would lead to a more widespread acceptance of GM crops in that country. Instead, in November 2002, India refused food-aid shipments of corn and soybeans from the United States on the grounds that they might contain genetically modified material (Guterl and Hardin, 2003). And China—ironically the first nation worldwide to plant a GM crop commercially (virus-resistant tobacco in 1989)—has recently developed a far more precautionary stance toward GM crops, not just regarding imports from the United States and elsewhere, but even toward its own farmers (Guterl and Hardin, 2003). Will the Gene Revolution, which began to take effect in these nations, come to a halt there as well?

This question will be answered in part by how the political scene plays out between the United States and the EU regarding trade of GM crops and food. As of this writing, the United States, along with Canada and Australia, has been involved in formally challenging EU regulations toward GM foods and crops through the dispute settlement body of the World Trade Organization (WTO). The United States has a reasonably solid scientific and legal case for this challenge, because WTO's Sanitary and Phytosanitary Measures state that nations may ban imports only on the condition of scientific evidence of risk. Thus far, 81 separate scientific studies financed by the European Commission have all shown no evidence of risk to health or the environment from GMOs (Paarlberg, 2003).

Other Crucial Differences in the Political Worlds of the Green and Gene Revolutions

Two other important differences in the political worlds of the Green and Gene Revolutions involve nongovernmental organizations (NGOs), public awareness, and public opinion. Unlike NGOs at the time of the Green Revolution, today's NGOs worldwide are an important political voice in the success or failure of agricultural revolutions. It was not until the 1970s and 1980s that NGOs representing environmental concerns gained full force (Shabecoff, 2000). At least part of the motivation for the agri-environmental NGO movement stemmed from the drawbacks of the Green Revolution with its myriad pesticides and fertilizers that polluted local soil, water, and air. Some of these environmental problems stemming from post-World War II agricultural advances were described in Rachel Carson's seminal work *Silent Spring*, first published in 1962.

In the realm of the debate over the Gene Revolution in today's political climate, many NGOs have aligned themselves against worldwide adoption of GM crops. Since 1998, NGOs have been instrumental in forcing the EU to impose a moratorium on new GM crop approvals, and now they are working to prevent approvals in the developing world. Greenpeace in particular has invested $7 million to halt planting of GM crops, especially in developing nations that have not yet planted GM crops (Paarlberg, 2003).

Furthermore, the public today is much more aware of what is going on in the agricultural world and food markets compared with the public during the days of the Green Revolution. It is not merely that genetically modified crops are more controversial, but also new information technologies (such as the World Wide Web) have greatly increased public access to coverage of such issues. Public surveys on the topic of genetically modified foods have helped to influence decisionmaking within the EU—in particular, surveys that revealed the desire on the part of citizens for the labeling of GM foods. Indeed, the public information situation is dramatically different from the situation during the heyday of the Green Revolution, and it could prevent the Gene Revolution from having the same impact as the Green Revolution.

The public opinion problem may be due to the fact that certain types of GM crops that were first introduced seemed to have no noticeable benefit to the general public. Taking Bt corn in the United States as an example, as shown in Table 3.3, consumer surplus (i.e.,

Table 3.3
Impact of Bt Corn on U.S. Stakeholders

Stakeholder Group	Benefits of Growing Bt Corn	Total Expected Gains to Stakeholder Groups ($ millions)	Impact on Welfare
Bt corn growers	Profit from increased yield Reduced pesticide costs Reduced farm worker illnesses from pesticides Improved corn quality for greater marketability	190 (–33.3 to 822)	12% increase in revenues
Non-Bt corn growers	Negative benefit (net loss) from reduced corn price	–416 (0 to 960)	6.7% loss in revenues
Consumers	Reduced corn prices Improved environmental quality	530 (0 to 1,200)	$1.90/ person per year
Agricultural biotechnology industry	Bt technology fee on seed paid by farmers using the seed	128 (96 to 160)	Owners of firms benefit from technology fees
Total welfare gain		432 (63 to 1290)	

SOURCE: Wu, 2004.
NOTE: Numbers in parentheses are 95% confidence intervals.

the gain to consumers) through reduced market price of corn and corn products represents the largest total public welfare increase among stakeholders—-about $530 million per year—a larger gain to consumers than to Bt corn growers as a whole. On a per-person basis, however, the consumer benefit amounts to less than $2 per year. Although consumers are price responsive and genetically modified corn can result in overall lower average consumer prices for corn, the savings are not large enough to offset consumers' perceived risks and their rejection of GM crops.

The political scene of the Gene Revolution has thus far proven to be dramatically different from that of the Green Revolution. The world is more complicated and stakeholders are more closely integrated. Food regulations in one part of the world could have a dramatic and often unintended impact on agricultural systems in another part of the world. The acceptance and even encouragement of new technologies that accompanied the Green Revolution are much more scarce in the Gene Revolution. Governments as well as consumers must accept GM crop technology if a Gene Revolution is to occur. Otherwise, there is serious doubt about how far the GM crop movement can spread worldwide.

In the next chapter, we address certain issues that the Green Revolution left unresolved and mistakes that the GM crop movement should avoid.

CHAPTER FOUR

Lessons for the Gene Revolution from the Green Revolution

Genetically modified crop technology has revolutionized agriculture in the United States, Canada, China, and Argentina. It exhibits the potential to have much wider impact, solving many of the current problems in agriculture worldwide. The types of GM crops that may become available in the future could boost crop yields while enhancing the nutritional value of staple foods and eliminating the need for inputs that could be harmful to the environment. While the environmental, health, and economic risks of GM crops should be carefully studied before full-scale adoption, the types of GM crops that are already available have thus far largely proven to be beneficial to agriculture and even to the environment, without evidence of adverse health or environmental impacts.

Yet, in other than the four countries mentioned above, the GM crop movement has had little or no impact. In those parts of the developing world where an agricultural revolution might be most welcome, the Gene Revolution has yet to be embraced. Why is this so?

For one thing, the Gene Revolution began in a different way than the Green Revolution. GM crops were first created within the context of the biotechnology industry to provide enhanced agricultural technologies to the industry's primary customers—farmers in the industrial world. These crops were not meant at the outset to be a life-saving technology for the developing world. Although it is almost certainly possible from a scientific and technological standpoint to create GM crops that would be beneficial to developing-world farmers, neither producers (the biotech industry) nor consumers (developing-world farmers) have sufficient economic incentives for this to happen. In fact, the enormous costs of producing each GM crop variety could prove to be a disincentive for the industry to develop "orphan GM crops" that would benefit developing-world farmers.

Additionally, even if the biotech industry were to develop GM crops that are beneficial to farmers in the developing world, the poorest of those farmers would not be able to afford GM crop seed instead of conventional varieties, much less purchase new GM crop seed for every planting season, as biotech patents would require them to do.

Finally, the current political situation is not as conducive to promoting this new agricultural movement as it was for the Green Revolution. For all the potential that GM technology holds, there are many challenges to be overcome if GM crops are to truly introduce a "Gene Revolution" worldwide.

Agricultural Biotechnology Is Just One of Several Options for the Future

Given the challenges stated above, it is important to keep in mind that agricultural biotechnology may not be the best solution, or even a one-shot solution, for all parts of the developing world, for three reasons.

First, as of yet, there are few if any sustainable technological solutions for controlling pests and pathogens in subtropical subsistence agriculture. Currently, in the poorest agricultural areas, food production is feasible only with very low inputs of semi-landrace material[1] of many different genotypes planted together to be broadly adapted to local environments. If one genotype fails, then the others may still succeed on a year-to-year basis, thereby achieving some level of security in the food supply (Miller, 2002). GM crops, unless they are cre-

---

[1] A "landrace" is a crop strain that is developed in traditional, typically poorer agricultural settings in the following way: Farmers note which crops seem to do best in their fields under various conditions and carefully set aside the seeds from those crops to sow the following year. Over time, there will be several landraces that are good under certain conditions. Semi-landrace simply refers to a less-careful job of selection.

ated from many different hybrids and are modified to withstand a broad range of environmental fluctuations, could not be expected to consistently improve yield if planted alone in subtropical areas.

Second, there are usually alternative ways to conduct public health or agricultural interventions, and all interventions have attendant costs. GM crops may be among the more costly interventions given their current R&D costs as well as the costs to growers. Malnourished people may not need GM golden rice to prevent blindness, for example, and policymakers should first take a step back to see which choices make the most sense in terms of both long-term sustainability and cost considerations. One possible intervention is enhanced conventional breeding. The newest conventionally bred crops have some immunity to common plant diseases and resistance to pests while retaining high yields (Pinstrup-Andersen and Schioler, 2001). Conventional breeding, while theoretically having far greater limitations than agricultural biotechnology, is less controversial from a global viewpoint and may be less expensive. Hence, in the short term, enhanced conventional breeding may be crucial to improving agricultural yields in areas that do not want to risk losing their food export markets due to current political tensions or government regulations, and it may be important to farmers with limited monetary resources. Other methods of promoting sustainable agriculture may also prove to be useful—for example, the adoption of farming techniques for greater economic return, such as agroforestry (to increase income), reclamation of degraded land, and irrigation scheduling (Pretty, 2003). As an alternative to introducing GM seeds now, a possible intervention that could be helpful in the poorest nations is the empowerment of women, who are currently the crop harvesters (Miller, 2002). For example, they could be educated to become agricultural scientists, learning to select seeds for desirable qualities, such as improved yield and improved quality. This could be a first step toward agricultural independence, which could then make for a smoother transition to agricultural commercialization.

Third, it would be overly simplistic to imagine that improved crop varieties, whether GM or enhanced conventional crops, are all that are needed to ensure food security. It is important to remember that the root cause of hunger is poverty—the inability to access food or the lack of a means to produce it (Hardin, 2003). Many factors contribute to poverty, not just poor food production. Farmers in the developing world also need support from certain political and social infrastructures that can safeguard incentives to use the GM crop technology appropriately. If the Gene Revolution is to succeed in the developing world, many of those infrastructures must be in place to ensure the long-term benefits from GM crop planting.

Broadening the Impact of the GM Crop Movement: Applying Lessons from the Green Revolution

Notwithstanding its attendant challenges and alternatives, the GM crop movement shows great promise. Like the Green Revolution before it, the GM crop movement has the potential to achieve substantial production increases in regions of need and (unlike the Green Revolution) to reduce the need for agricultural chemicals and scarce resources, such as water. Both the successes and failures of the Green Revolution provide useful lessons for how to make GM crop technology a desirable and sustainable agricultural movement in the developing world.

The Green Revolution demonstrates that to create GM crops that are truly beneficial to the developing world, plant breeders and other scientists must be familiar with the local environment and the planting methods of the region for which they are developing crops. Oftentimes, agricultural conditions in developing regions are so different from those in the industrial world that it is difficult for industrial-world scientists to know how to devise appropriate technologies for those regions. During the Green Revolution, plant scientists traveled abroad extensively, developing crop seeds that were best suited to particular regions given their particular weather conditions, plant pests, water availability, and planting seasons. Importantly, these plant scientists trained others in each region to be able to carry out the Green Revolution practices independently. The same sort of global effort is needed for the Gene Revolution to take hold in the developing world.

For this global effort to take place, however, there must be a vested interest on the part of those entities that control the Gene Revolution technologies—those that create the technologies, namely those in the biotech industry and those that regulate the technologies nationally and internationally. The Green Revolution owed much of its success to public-sector institutions that poured resources into the effort, as well as to regulatory regimes in both the donor and recipient nations that were permissive and even encouraged adoption of the new agricultural technologies. Times have changed, though. R&D for GM crops is supported by the public sector only in unusual circumstances, with the biotechnology industry mostly creating GM crops that are beneficial to industrial-world farmers, its primary customers (and those who can afford to pay for the technology).

To complicate matters, current intellectual property regulations that protect the biotech industry's creations limit the flow of information on how to create GM seeds to the public sector, making it difficult to garner the public support needed to develop crops for the poorest farmers in the world. IP rights also lead indirectly to increased GM seed costs that make GM seeds unaffordable to most developing-world farmers without significant subsidization. Collaborations between the public and the private sectors to promote the Gene Revolution in the developing world do exist, but thus far only in isolated instances on a small scale.

Further hindering GM crop adoption worldwide is the lack of uniform regulation of foods derived from modern biotechnology. Unlike the permissive regulatory environment of the Green Revolution, in which agricultural advances were encouraged for both philanthropic and political reasons, decisionmakers today are largely divided into two camps on whether GM crops should flow freely through the food system. The European Union's new regulations on traceability and labeling of GM foods would require a crop-segregation system that is almost impossible to achieve in a nation without a highly developed commercial agricultural sector.

Thus, developing nations may find it in their best interest to avoid planting GM crops altogether, despite the agricultural and nutritional benefits that GM crops might provide. In addition, many NGOs and other organizations have expressed concerns about the risks surrounding GM crops, and their opinions are becoming increasingly important to the public debate and decisionmaking process. These groups and the average citizen have seen little public benefit from the types of GM crops produced today, except for perhaps slightly cheaper food (Wu, 2004).

What can we determine about the prospects for the Gene Revolution by studying the Green Revolution's successes and failures? The Gene Revolution thus far resembles the Green Revolution in the following ways: (1) It employs new science and technology to create crop seeds that can significantly outperform the types of seeds that preceded it; (2) the impact of the new seed technologies can be critically important to developing-world agriculture; and (3) for a variety of reasons, these technologies have not yet reached parts of the world where they could be most beneficial. On the other hand, the Gene Revolution is unlike the Green Revolution in the following ways: (1) The science and technology required to create GM crop seeds are far more complicated than the science and technology used to create Green Revolution agricultural advancements; (2) GM seeds are created largely through private enterprise rather than through public-sector efforts; and (3) the political climate in which agricultural science can introduce new technologies has changed dramatically.

The similarities and differences between the Green and Gene Revolutions lead us to speculate that for the GM crop movement to have the sort of impact that would constitute an agricultural revolution, the following goals still need to be met and their related challenges overcome:

1. **Agricultural biotechnology must be made affordable to developing-world farmers.** Unless this condition is met, farmers may not see that it is in their best interest to use GM crops, despite the significant benefits those crops could provide.

During the Green Revolution, the new HYV seeds and accompanying chemicals were more expensive than the landrace seeds that developing-world farmers typically had used. Therefore, loan systems and cost-reduction programs were established regionally in which farmers' eventual profits from increased production could be used to reimburse lenders. In many settings, these programs proved to be no longer necessary several years after their successful adoption. Current R&D costs for genetically modified seeds are even higher than the R&D costs for the Green Revolution's HYV seeds. At the price that U.S. farmers currently pay, GM seeds would be unaffordable to most developing-world farmers. Cost-reduction programs and loan systems similar to those that were established during the Green Revolution must also be established for the Gene Revolution; however, establishing such systems is more difficult now because of higher costs and because the seeds are produced by the biotech industry rather than by agricultural scientists in the public sector.

2. There is a need for larger investments in research in the public sector. Numerous studies (e.g., Alston et al., 1995; Conway, 1998; Shoemaker et al., 2001) have shown the importance of public-sector R&D to agricultural advancements, including the advancements of the Green Revolution. During the Green Revolution, partly because the R&D and its products were almost entirely in the public domain, intellectual property issues were not a barrier to scientists, for example, taking seeds from one region of the world, hybridizing them with seeds from another region, and producing new seeds to benefit yet another region. Today, however, the production and distribution of GM crops are largely within the domain of the biotech industry, and IP issues are central to the development of GM seed. While IP laws protect the rights of GM seed creators in industry, those laws are currently an impediment to disseminating the necessary knowledge and technology to those parts of the world that need them. Therefore, public-sector research is essential if the GM movement is to assume revolutionary proportions. Partnerships between the public and private sectors can result in the more efficient production of GM crops that are useful to the developing world and expand the accessibility of those crops and their associated technologies to developing-world farmers.

3. To garner the level of public interest and support that can sustain an agricultural revolution, agricultural development must be regarded as being critically important from a policy perspective, in both donor and recipient nations. Without public policy support, cooperation among the many stakeholders in the Gene Revolution will be stymied.

For 30 years after World War II, policymakers viewed agricultural development as being essential to world peace. For that reason, policymakers in both the United States and in Asia and Latin America supported the Green Revolution from the start. The end of the Cold War, however, has not brought about an increase in global stability. Whereas the conflict between East and West has declined, there is a growing divide between rich and poor nations. Unfortunately, with the end of the Cold War, developed nations are concentrating more closely on their domestic political agendas and less on global concerns, and as such have decreased their funding to poorer nations. However, these reductions in aid are not in the best long-term interests of even industrialized nations. An increasingly polarized world of the rich versus the poor will result in growing political unrest. Unless developing nations are helped to provide sufficient food, employment, and shelter for their growing populations, the political stability of the world will be further undermined (Conway, 1998).

As population numbers continue to increase, agricultural development is more necessary than ever to eliminate malnutrition and prevent famine, particularly in sub-Saharan Africa. GM crops are seen as a means for addressing those problems. However, policymakers worldwide are far from being a combined force on this issue; the driving force behind improved agriculture is less unified than it was during the Green Revolution. The question of who should assume the task of re-establishing the importance of agricultural development among policymakers is an issue for further inquiry.

4. Policymakers in the developing world must set regulatory standards that take into consideration the risks as well as the benefits of foods derived from GM crops. This goal is crucial to the cooperation of the many stakeholders that are affected by GM crops

and also for the sustainability of the GM crop movement in the fore-seeable future. A generation ago, the regulatory environment sur-rounding the Green Revolution was extremely permissive. Scientists could move freely among nations to help breed and plant HYV crops, and there was no stigma attached to eating foods developed from these crops. Today, however, the regulatory world is divided between those nations that permit GM crops to move freely through their food system (e.g., the United States, Canada, China, and Argentina) and those (primarily the EU) that have strict regulations regarding GM crops in their food systems. There are many possible reasons for the disparity in regulations—differing consumer attitudes, trade is-sues, and differences in regulatory philosophy among them.

The discord regarding GM crop regulations is currently playing itself out (as of this writing) in a case before the WTO to determine whether the EU's rules on GM foods constitute an illegal trade bar-rier. In the meantime, policymakers in certain African nations have decided that they cannot afford to permit GM crop planting, even if it is beneficial to their growers and consumers, because they are wary of losing financial aid from the EU if they are seen as taking a pro-GM crop stance. Without regulations that explicitly take into ac-count potential benefits to both farmers and consumers, those nations that might stand to benefit most from GM crops may be discouraged from allowing them to be planted.

At the same time, policymakers worldwide must ensure that risk assessments of GM crops are conducted to address the specific con-cerns of their regions. A risk assessment of transgene outflow in the United States, for example, is unlikely to be relevant to ecological concerns in Mexico or Africa. In assessing risks, policymakers in de-veloping nations must consider, among other factors, the types of na-tive and agricultural plants that may be affected by the presence of GM crops, traditional farming practices, and the desired traits of GM crops that may be planted in their regions in the near term and long term.

Implications for Relevant Stakeholders

What do these challenges mean for the various stakeholders that are or should be involved in solving the problems surrounding current and future agricultural needs worldwide? We offer recommendations for four stakeholder groups: the U.S. government and other national governments worldwide, public institutions, private companies, and NGOs.

First, national governments worldwide should realize that so long as there is any threat of widespread hunger or malnutrition, the threat of political instability and insecurity (partly caused by lack of food security) is larger. Indeed, problems of hunger and malnutrition still exist, most especially in sub-Saharan Africa, and the benefits of the Green Revolution in other parts of the developing world are slowing. Thus, governments should pay closer attention and lend greater support to agriculture and food policies regarding developing nations in need.

Public institutions—foundations, agricultural departments in universities, and other national and international agricultural research organizations—should have this same sort of realization when planning their agendas and areas of focus. In addition to the national security issues, they must recognize the problem of continuing hunger and malnutrition as an important public welfare problem.

From a technological standpoint, private companies are in a position of power because they possess the scientific knowledge and capabilities to produce GM crop seeds that could have significant benefit worldwide. However, unless companies use that power for global good, their products (i.e., GM crop seeds) may continue to be stigmatized in many parts of the world, with serious market implications. Therefore, private companies should use their technological know-how to focus on the needs of developing-world farmers and should partner with public institutions to benefit from a mutual sharing of resources.

Nongovernmental organizations should strive to present more-balanced perspectives on the GM crop issue, keeping in mind their increased level of influence (and corresponding responsibility) in recent years regarding policy decisions on adoption of new technologies. NGOs that support the GM crop movement should make it clear that not all the potential risks of agricultural biotechnology have been researched. NGOs that are against GM crops should not mislead the public about any risks that have already been proven to be insignificant, nor decline to spread the message about potential benefits from GM crops. All NGOs should help to communicate the message that the risks associated with planting certain types of GM crops in specific locales worldwide should be carefully considered.

The challenges discussed in this chapter are interrelated. Revised regulations on genetically modified crops must accompany widespread collective policy efforts to revitalize agricultural development. And before developing-world farmers and consumers can benefit from GM crops or any other type of enhanced crop breeding, the technologies must be affordable and farmers must understand how to use them. The GM crop movement must overcome an intertwined collection of challenges before it can have an impact beyond those regions of the world that already produce excesses of food. If the GM crop movement can overcome these challenges, while proving itself to be acceptably free of adverse health and environmental impacts, it has the potential to provide benefits to farmers and consumers around the globe in previously inconceivable ways, while mitigating the need to use potentially harmful chemicals or scarce water supplies for agriculture. It can then, indeed, become a true "Gene Revolution."

Bibliography

51 *Federal Register* at 23302, 1986.

57 *Federal Register* at 22991, 1992.

66 *Federal Register* at 4706, 2001.

Alston, Julian M., George W. Norton, and Philip G. Pardey, *Science Under Scarcity: Principles and Practice for Agricultural Research Evaluation and Priority Setting,* Ithaca, N.Y.: Cornell University Press, 1995.

Andow, D. A., and D. N. Alstad, "F2 Screen for Rare Resistant Alleles," *Journal of Economic Entomology,* Vol. 91, 1998, pp. 572–578.

Baum, Warren C., *Partners Against Hunger: The Consultative Group on International Agricultural Research,* Washington, D.C.: World Bank, 1986.

Borlaug, Norman, "The Green Revolution: Its Origins and Contributions to World Agriculture," speech given at the School of Agriculture, Purdue University, West Lafayette, Ind., February 7, 2003.

Boserup, Ester, *The Conditions of Agricultural Growth: The Economics of Agrarian Change Under Population Pressure,* Chicago: Aldine, 1965.

Brea, Jorge, "Population Dynamics in Latin America," *Population Bulletin,* Vol. 58, No. 1, Washington, D.C.: Population Reference Bureau, 2003.

Byrne, Pat, Sarah Ward, Judy Harrington, and Lucy Fuller, "Transgenic Crops: An Introduction and Resource Guide," Department of Soil and Crop Sciences, Colorado State University, 2004 (available at http://www.colostate.edu/programs/lifesciences/TransgenicCrops as of April 2004).

Carpenter, Janet, *Case Studies in Benefits and Risks of Agricultural Biotechnology: Roundup Ready Soybeans and Bt Field Corn,* Washington, D.C.: National Center for Food and Agricultural Policy Report, 2001.

Carson, Rachel, *Silent Spring,* New York: Houghton Mifflin Company, 1962 (Mariner Books edition, 2002).

Cayford, Jerry, "Breeding Sanity into the GM Food Debate," *Issues in Science and Technology,* Winter 2004, pp. 49–56.

Chandler, Robert E., *A History of IRRI,* Metro Manila, Philippines: International Rice Research Institute, 1982.

Cleaver, Harry, "The Contradictions of the Green Revolution," *American Economic Review,* Vol. 62, Nos. 1 and 2, 1972, pp. 177–186.

Conway, Gordon, *The Doubly Green Revolution: Food for All in the 21st Century,* Ithaca, N.Y.: Cornell University Press, 1998.

Conway, Gordon, "Biotechnology and Hunger," remarks addressed to Senate about Science at the House of Lords, London, UK, May 8, 2003.

Dowd, Patrick F., "Biotic and Abiotic Factors Limiting Efficacy of Bt Corn in Indirectly Reducing Mycotoxin Levels in Commercial Fields," *Journal of Economic Entomology,* Vol. 94, No. 5, 2001, pp. 1067–1074.

Echols, Marsha, "Caught in the Middle? The Impact of the Biotechnology Debate on Africa and Other Developing Countries," *Proceedings of Ceres Roundtable: Breaking the EU/U.S. Deadlock on Foods Derived from Modern Biotechnology: A Global Perspective,* Center for Food and Nutrition Policy, Blacksburg, Va.: Virginia Tech, 2002.

European Economic Community, "Council Directive 90/220 on the Deliberate Release to the Environment of Genetically Modified Organisms," *Official Journal of the European Communities,* Vol. 117, No. 15, May 8, 1990.

Farmer, B. H., "Perspectives on the Green Revolution in South Asia," *Modern Asian Studies,* Vol. 20, No. 1, 1986, pp. 175–199.

Federal Insecticide, Fungicide, and Rodenticide Act (7 U.S.C. Section 136, 1972).

Gould, F., A. Anderson, D. Summerford, D. Heckel, J. Lopez, S. Micinski, R. Leonard, and M. Laster, "Initial Frequency of Alleles for Resistance to Bacillus thuringiensis Toxins in Field Populations of Heliothis virescens," *Proceedings of the National Academy of Sciences,* Vol. 94, 1997, pp. 3519–3523.

Griliches, Zvi, "Hybrid Corn: An Exploration in the Economics of Technological Change," *Econometrica,* Vol. 25, 1957, pp. 501–522.

Guterl, Fred, and Lowell S. Hardin, "The Fear of Food," *Newsweek International,* January 27, 2003.

Hammond, B., K. Campbell, C. Pilcher, A. Robinson, D. Melcion, B. Cahagnier, J. Richard, J. Sequeira, J. Cea, F. Tatli, R. Grogna, A. Pietri, G. Piva, and L. Rice, "Reduction of fumonisin mycotoxins in Bt Corn," *The Toxicologist,* Vol. 72, 2003, p. S-1.

Hardin, Lowell, Purdue University, personal communications, December 2003.

Harriss, John, "The Green Revolution in North Arcot: Economic Trends, Household Mobility, and the Politics of an 'Awkward Class,'" in Peter B. R. Hazell and C. Ramasamy, *The Green Revolution Reconsidered: The Impact of High-Yielding Rice Varieties in South India,* Baltimore, Md.: The Johns Hopkins University Press, 1991.

Hazell, Peter B. R., and C. Ramasamy, *The Green Revolution Reconsidered: The Impact of High-Yielding Rice Varieties in South India,* Baltimore, Md.: The Johns Hopkins University Press, 1991.

Homer-Dixon, Thomas F., "On the Threshold: Environmental Changes as Causes of Acute Conflict," *International Security,* Vol. 16, No. 2, 1991, pp. 76–116.

James, Clive, *Global Statutes of Commercialized Transgenic Crops,* Ithaca, N.Y.: International Service for the Acquisition of Agri-biotech Applications (ISAAA), ISAAA Briefs No. 30, 2003.

James, Clive, *Progressing Public-Private Sector Partnerships in International Agricultural Research and Development,* Ithaca, N.Y.: ISAAA, ISAAA Briefs No. 4, 1997.

Khush, Gurdev S., "Green Revolution: the Way Forward," *Nature Reviews: Genetics,* Vol. 2, 2001, pp. 815–821.

Kilman, Scott, and Roger Thurow, "Africa Could Feed Itself but Many Ask: Should It?" *Wall Street Journal,* December 3, 2002, p. 1.

King, Neil, "U.S. Ponders Next Course in EU Biotech-Food Fight," Politics and Policy, *Wall Street Journal,* December 2, 2002.

Laget, Patrice, and Mark Cantley, "European Responses to Biotechnology: Research, Regulation, and Dialogue," *Issues in Science and Technology,* Vol. 17, No. 4, 2001, pp. 37–42.

Levidow, Les, "Regulating Bt Maize in the United States and Europe: A Scientific-Cultural Comparison," *Environment,* Vol. 41, No. 10, 1999, pp. 10–22.

Lin, W., W. Chambers, and J. Harwood, *Biotechnology: U.S. Grain Handlers Look Ahead,* Washington, D.C.: Economic Research Service–USDA, Agricultural Outlook AGO-270, 2000, pp. 29–34.

Lipton, Michael, "Growing Mountain, Shrinking Mouse? Indian Poverty and British Bilateral Aid," *Modern Asian Studies,* Vol. 30, No. 3, 1996, pp. 481–522.

Lipton, Michael, and Richard Longhurst, *New Seeds and Poor People,* London, UK: Unwin Hyman, 1989.

Löfstedt, Ragnar E., and David Vogel, "The Changing Character of Regulation: A Comparison of Europe and the United States," *Risk Analysis,* Vol. 21, No. 3, 2001, pp. 399–405.

Marris, C., B. Wynne, P. Simmons, and S. Weldon, *Public Perceptions of Agricultural Biotechnologies in Europe,* Final Report of the PABE Research Project, Commission of European Communities, FAIR CT98-3844 (DG12-SSMI), 2002 (available at http://www.lancs.ac.uk/depts/ieppp/pabe/docs/pabe_finalreport.pdf as of May 2004).

Miller, J. David, Carleton University, Ottawa, Ont., personal communications, September 24, 2002, and October 14, 2002.

Munkvold, Gary P., and Richard L. Hellmich, "Comparison of Fumonisin Concentrations in Kernels of Transgenic Bt Maize Hybrids and Non-transgenic Hybrids," *Plant Disease,* Vol. 83, No. 2, 1999, pp. 130–138.

Nap, Jan-Peter, Peter L. J. Metz, Marga Escaler, and Anthony J. Conner, "The Release of Genetically Modified Crops into the Environment: Overview of Current Status and Regulations," *The Plant Journal,* Vol. 33, 2003, pp. 1–18.

National Academy of Sciences, *Genetically Modified Pest-Protected Plants: Science and Regulation,* Washington, D.C.: National Academy Press, 2000a.

National Academy of Sciences, *Transgenic Plants and World Agriculture,* Washington, D.C.: National Academy Press, 2000b.

Nelson, G. C., J. Babinard, and T. Josling, "The Domestic and Regional Regulatory Environment," in *Genetically Modified Organisms in Agriculture,* G. C. Nelson, ed., San Diego: Academic Press, 2001.

Official Journal, L 106, April, 17, 2001.

Orme, John, "The Utility of Force in a World of Scarcity," *International Security,* Vol. 22, No. 3, Winter 1997–1998, pp. 138–167.

Paarlberg, Robert, *The Politics of Precaution: Genetically Modified Crops in Developing Countries,* Baltimore, Md.: The Johns Hopkins University Press, 2001.

Paarlberg, Robert, "African Famine, Made in Europe," op-ed column, *Wall Street Journal,* August 23, 2002.

Paarlberg, Robert, "Reinvigorating Genetically Modified Crops," *Issues in Science and Technology,* Spring 2003.

Perkins, John H., *Geopolitics and the Green Revolution: Wheat, Genes, and the Cold War,* New York: Oxford University Press, 1997.

Pingali, Prabhu L., and Paul W. Heisey, "Cereal Crop Productivity in Developing Countries: Past Trends and Future Prospects," in *Agricultural Science Policy,* Julian M. Alston, Philip G. Pardey, and Michael J. Taylor, eds., Baltimore: The Johns Hopkins University Press, 2001.

Pinstrup-Andersen, Per, and M. Jaramillo, "The Impact of Technological Change in Rice Production on Food Consumption and Nutrition," in Peter B. R. Hazell and C. Ramasamy, *The Green Revolution Reconsidered: The Impact of High-Yielding Rice Varieties in South India,* Baltimore, Md.: The Johns Hopkins University Press, 1991.

Pinstrup-Andersen, Per, and Ebbe Schioler, *Seeds of Contention,* Baltimore: The Johns Hopkins University Press, 2001.

Pretty, Jules, "Agroecology in Developing Countries: The Promise of a Sustainable Harvest," *Environment,* Vol. 45, No. 9, 2003, pp. 8–20.

Qaim, Matin, and David Zilberman, "Yield Effects of Genetically Modified Crops in Developing Countries," *Science,* Vol. 299, 2003, pp. 900–902.

Ricardo, David, *On the Principles of Political Economy and Taxation,* 3rd edition, 1821 (in print, New York: George Olms Publishers, 1977).

Ruttan, Vernon, "Foreword to Cornell Paperbacks Edition," in G. Conway, *The Doubly Green Revolution: Food for All in the 21st Century*, Ithaca, N.Y.: Cornell University Press, 1998.

Sachs, Jeffrey, "Geography and Economic Transition," working paper, Harvard Institute for International Development, November 1997 (available at http://www2.cid.harvard.edu/hiidpapers/geotrans.pdf as of April 2004).

Schaafsma, A. W., D. C. Hooker, T. S. Baute, and L. Illincic-Tamburic, "Effect of Bt-Corn Hybrids on Deoxynivalenol Content in Grain at Harvest," *Plant Disease*, Vol. 86, No. 10, 2002, pp. 1123–1126,

Sears, Mark K., Richard L. Hellmich, Diane E. Stanley-Horn, Karen S. Oberhauser, John M. Pleasants, Heather R. Mattila, Blair D. Siegfried, and Galen P. Dively, "Impact of Bt Corn Pollen on Monarch Butterfly Populations: A Risk Assessment," *PNAS Online,* September 14, 2001 (available at www.pnas.org/cgi/doi/10.1073/pnas.211329998 as of April 2004).

Serageldin, Ismail, "Foreword to UK Edition," in G. Conway, *The Doubly Green Revolution: Food for All in the 21st Century*, Ithaca, N.Y.: Cornell University Press, 1998.

Shabecoff, Philip, *Earth Rising: American Environmentalism in the 21st Century,* Washington, D.C.: Island Press, 2000.

Shoemaker, Robbin, Joy Harwood, Kelly Day-Rubenstein, Terri Dunahay, Paul Heisey, Linwood Hoffman, Cassandra Klotz-Ingram, William Lin, Lorraine Mitchell, William McBride, and Jorge Fernandez-Cornejo, "Economic Issues in Agricultural Biotechnology," *ERS Agriculture Information Bulletin No. 762,* Washington, D.C.: Economic Research Service, March 2001.

Toenniessen, Gary H., "Vitamin A Deficiency and Golden Rice: The Role of the Rockefeller Foundation," New York: The Rockefeller Foundation, November 14, 2000 (available at http://www.rockfound.org/display.asp?context=1&Collection=4&DocID=80&Preview=0&ARCurrent=1 as of April 2004).

Truman, Harry S., "Inaugural Address," January 20, 1949, from the Page By Page Books website (http://www.pagebypagebooks.com/Harry_S_Truman/Inaugural_Address/Inaugural_Address_p1.html).

United Nations Children's Fund, *The State of the World's Children 1998,* New York: Oxford University Press for UNICEF, 1998.

United Nations Development Programme, *Human Development Report, 1994,* New York: UNDP, 1994.

United Nations Population Division, "World Population Prospects, The 2002 Revision Population Database," 2002 (available at http://esa.un.org/unpp as of April 2004).

U.S. Congress 1958, Section 402(a)(1).

U.S. Department of Agriculture, Animal and Plant Health Inspection Service, Wildlife Services, "A Historical Overview of the U.S. Department of Agriculture, National Wildlife Research Center's International Research Programs," n.d. (available at http://www.aphis.usda.gov/ws/nwrc/hx/international.html as of April 2004).

U.S. Department of Agriculture, National Agricultural Statistics Service, *Historical Track Records,* May 2001 (available at http://www.usda.gov/nass/pubs/trackrec/trackrec2001.pdf as of April 2004).

U.S. Department of Agriculture, National Agricultural Statistics Service, "Prospective Plantings," Washington, D.C.: USDA, March 31, 2003 (available at http://usda.mannlib.cornell.edu/reports/nassr/field/pcp-bbp/pspl0303.pdf as of April 2004).

U.S. Environmental Protection Agency, *Biopesticides Registration Action Document: Bacillus thuringiensis Plant-Incorporated Protectants,* Environmental Protection Agency, 2001 (available at http://www.epa.gov/pesticides/biopesticides/pips/bt_brad.htm as of May 2004).

Woodward, John, Biometrics Management Office, personal communication, May 2003.

Wu, Felicia, "Tools for Regulatory Decisions Concerning Genetically Modified Corn," Ph.D. dissertation, Carnegie Mellon University, Pittsburgh, Pa., 2002.

Wu, Felicia, "Explaining Public Resistance to Genetically Modified Corn: An Analysis of the Benefits and Risks," *Risk Analysis,* Best Paper Issue, Vol. 24, No. 3, 2004, pp. 717–728.

Wu, Felicia, J. David Miller, and Elizabeth A. Casman, "The Economic Impact of Bt Corn Resulting from Mycotoxin Reduction," *Journal of Toxicology Toxin Reviews,* forthcoming.

Zhang, Hong-Xia, and Eduardo Blumwald, "Transgenic Salt-Tolerant Tomato Plants Accumulate Aalt in Foliage but Not in Fruit," *Nature Biotechnology,* Vol. 19, No. 8, 2001, pp. 765–768.